Eco-Literate Music Pedagogy

Eco-Literate Music Pedagogy examines the capacity of musicking to cultivate ecological literacy, approaching eco-literate music pedagogy through philosophical and autoethnographical lenses. Building on the principle that music contributes uniquely to human ecological thinking, this volume tracks the course of eco-literate music pedagogy while guiding the discussion forward:

- What does it mean to embrace the impulse to teach music *for* ecological literacy?
- What is it like to theorize eco-literate music pedagogy?
- What is learned through enacting this pedagogy?
- How do the impulsion, the theorizing, and the enacting relate to one another?

Music education for ecological consciousness is experienced in local places, and this study explores the theory underlying eco-literate music pedagogy in juxtaposition with the author's personal experiences. The work arrives at a new philosophy for music education: a spiritual praxis rooted in soil communities, one informed by ecology's intrinsic value for non-human being and musicking. *Eco-Literate Music Pedagogy* adds to the emerging body of music education literature considering ecological and environmental issues.

Daniel J. Shevock is a Lecturer in Music at Pennsylvania State University, Altoona, where he served as Emerging Musical Artist in Residence in Jazz. He taught instrumental and general music in the Pittsburgh Public Schools for eleven years.

Routledge New Directions in Music Education Series
Series Editor: Clint Randles

The **Routledge New Directions in Music Education Series** consists of concise monographs that attempt to bring more of the wide world of music, education, and society into the discourse in music education.

Eco-Literate Music Pedagogy
Daniel J. Shevock

The Music Profiles Learning Project
Radio Cremata, Gareth Dylan Smith, Joseph Michael Pignato, and Bryan Powell

Eco-Literate Music Pedagogy

Daniel J. Shevock
PENNSYLVANIA STATE UNIVERSITY, ALTOONA

Routledge
Taylor & Francis Group
NEW YORK AND LONDON

First published 2018
by Routledge
711 Third Avenue, New York, NY 10017

and by Routledge
2 Park Square, Milton Park, Abingdon, Oxon, OX14 4RN

Routledge is an imprint of the Taylor & Francis Group, an informa business

Library of Congress Cataloging in Publication Data
Names: Shevock, Daniel J., author.
Title: Eco-literate music pedagogy / Daniel J. Shevock.
Description: New York : Routledge, 2017. | Series: Routledge New Directions in Music Education Series
Identifiers: LCCN 2017016700 (print) | LCCN 2017021200 (ebook) | ISBN 9781315211596 (ebook) | ISBN 9780415792578 (hardback)
Subjects: LCSH: Ecomusicology. | Music—Instruction and study—Environmental aspects. | Environmentalism.
Classification: LCC ML3799.3 (ebook) | LCC ML3799.3.S54 2017 (print) | DDC 780.71—dc23
LC record available at https://lccn.loc.gov/2017016700

ISBN: 978-0-415-79257-8 (hbk)
ISBN: 978-1-315-21159-6 (ebk)

Typeset in Times New Roman
by Swales & Willis Ltd, Exeter, Devon, UK

Contents

Series Foreword

The *Routledge New Directions in Music Education Series* consists of concise monographs that attempt to bring more of the wide world of music, education, and society—and all of the conceptualizations and pragmatic implications that come with that world—into the discourse in music education. It is about discovering and uncovering big ideas for the profession, criticizing our long held assumptions, suggesting new courses of action, and putting ideas into motion for the prosperity of future generations of music makers, teachers of music, researchers, scholars, and society.

Clint Randles, Series Editor

Acknowledgements

Like reality, scholarship is ecological. This book is a product of interconnected relationships among beings, and not only the result of individual work. My wife and son (my hiking and gardening partners), Mercedes and Penny inspired this work by helping me recognize resonances—the musics of the spheres, of birds, wind, and soil. Vincent Bates's work on place, rural ideals, and sustainability in music education formed a basis for my scholarship. He read and commented extensively on every chapter of the book. Darrin Thornton, Valerie Flamini, and Joanne Rutkowski each read and commented on chapters. Clint Randles, the series editor, provided insight without which this work would have been impossible. The chickadees, perennial companions musicking at the backyard birdfeeder, improved this work. I'm grateful to these music educators, but any errors are my own.

Daniel J. Shevock

Prelude

> We stand now where two roads diverge. But unlike the roads in Robert Frost's familiar poem, they are not equally fair. The road we have long been travelling is deceptively easy, a smooth superhighway on which we progress with great speed, but at its end lies disaster. The other fork in the road—the one "less travelled by"—offers our last, our only chance to reach a destination that assures the preservation of our earth.
>
> (Rachel Carson, 1962[1])

Walking a Road Less Travelled By

We walk under a green canopy. For humans, walking as a mode of transportation and a meditation is natural. Its pace is natural, especially compared to flight or driving a car. Unlike motorized transport, there's no obvious machine mediating the walker from soil. Beneath my feet, I remain connected to soil, to dirt, to small plants, to Mother Earth, to life. Walking on pavement, a hiking trail, and grass each generate distinctive sounds. How should my sound, my life, blend with nature? Do my feet and heartbeat synchronize to the rhythms of nature?

Birds sing above, and insects below. Musicking without and within. Walkers can, and often do, talk with others, and convivial sounds become a part of an activity humankind has repeated for hundreds of thousands of years, since well before evolving into *homo sapiens*. These sounds make an inconsequential musicking activity, too often repeated to notice: except that music teachers and students do not often walk together. Certainly, band directors direct marching band students: "left, right, left, right." But just walking together is atypical, and simply *being* out of doors together is uncommon in music education.

On this particular July day, I was walking with youth chorale students during their yearly summer music camp at a park outside of rural West Milton, Pennsylvania. Like my hometown in Patton, this park contained green deciduous and coniferous trees, songbirds, squirrels, and deer. Both places are populated by rural people, town-folk, and farmers who modern industrial society ignores or forgets. Valerie invited me to guest-teach this session. Inspired by the eco-music concepts I was forming, I was re-pairing detached aspects of my life. I was joining my loves of hiking and music; combining countless hours spent toiling in practice rooms to childhood hours spent idly in Prince Gallitzin State Park; linking music education philosophy to ecological literacy. A fork in the road was turned on its head, converging to draw two parts of myself together. Of course, I was a little nervous about this session. Though I had thought about and theorized eco-literate music pedagogy, I had yet to try it in the real world.

As we walked and talked, my butterflies dissolved step by step in the delightful soundscape. Cicadas, sparrows, chickadees, cardinals, and the wind's crescendo through the leaves above soothed my fears, and I reflected on the activities I had planned—a listening meditation to feel this place's rhythm, echoing the songs in nature using singing voices, and co-composing music using these melodies and organic found sounds. By the time we arrived at Sylvan Chapel, a cement and brick altar with pews in the woods, I was in a calm state of mind, suitable for teaching and learning. I walked to the altar—am I a priest or teacher?—asked the students to close their eyes, to listen, and, in this simple, natural, ecological, *true* place, we took our first steps.

Introduction

Months before guest-lecturing, I went to the woods for inspiration. Walking in the woods, I desired coherence. Coherence among the wild, uncultivated musics of non-human animals and nature enveloping me since my youth, and my anthropocentric (human-centered) music teaching practice. Without knowing to where I would walk, my legs took me toward eco-literate music pedagogy. This walking began in February 2015, just after completing my doctoral dissertation. There is something freeing about completing such a large project. Free from work and in the habit of writing daily, walking became a jog and then a sprint. Inspired by deep ecology and ecofeminism, ecomusicology and ecological literacy, I wrote a short philosophical paper outlining possible directions for ecology in music education praxis (Shevock 2015b). Presenting this study at a conference, I met a music teacher who invited me to explore these concepts at a youth choir camp. This camp, which I have taught at twice at the time of writing, afforded a wonderful opportunity to share these ideas with the world outside of my own mind, where possibilities were transforming into practice. In this book, I reveal the eco-literate music pedagogy that I have attempted since beginning this walk. As an academic, I have attempted to include scholarship and personal experiences. While this book cannot hope to provide a full depiction of these experiences, it provides a map—an imperfect map, but also a longing, searching, and hopefully thoughtful map.

This Prelude introduces readers to the concept of eco-literate music pedagogy. The historical precedents for this concept began in time immemorial, and found institutional beginnings in the 1920s with Satis Coleman's once-popular creative music method. Her pedagogy drew from nature as a source of inspiration for classroom music activities. In the 1970s R. Murray Schafer's *soundscape* provided a rationale for the sounds of places becoming the object of study in music classrooms. However, neither of these historical precedents gained wide appeal in the music education profession. I have struggled with why this is—it may be due to music education's move from progressive to aesthetic ideals in the mid-20th century, or it may be an anthropocentric valuing of composers above the possibility of non-human musics. Perhaps the lack of an eco-literate music pedagogy is the result of an institutionalized avoidance of ecological realities, or perhaps a result of institutionalized patriarchy. Ecofeminists have drawn connections between the unjust dominations of women and nature. Activist Judi Bari (1994) wrote, "it is the hatred of feminine, which is the hatred of life, that has helped bring about the destruction of the planet. And it is the strength of women that can restore the balance we need to survive" (225). An awareness of music education as a patriarchal institution might help us reestablish Satis Coleman's insights, to live well in our places, and address ecological crises. As a patriarchal institution, music education has had a blind faith in unsustainable technological fixes, synchronized to Rachel Carson's insights into blind faith in DDT in her day.

More recent calls for environmental sustainability in music education by Vincent C. Bates, Charlene A. Morton, and Julia Eklund Koza[2] seem not to be widely heeded, leading to the current book's guiding questions. In previous work, I introduced music education *for* ecological consciousness and ecological literacy, including an essential structure (Shevock 2015b). I used this essential structure, shared later in this prelude, to make pedagogical choices. Reflective writing, a staple of environmentalism since Thoreau, is used to introduce autoethnographic writing in music education and the current method, philosophy/autoethnography. Though this composite term is unique to this book, philosophy/autoethnography makes explicit the personal nature of writing environmental philosophy. If postmodernism is generally understood as skepticism toward metanarratives, philosophy/autoethnography provides a subjective approach to theoretical questions in line with the postmodern and ecological challenges. I explain the method, which led to reflections, such as "Walking a Road Less Travelled By," and fictional "creative vignettes." Each chapter of this book begins with a reflection, includes a vignette, and ends with a poem. These represent the analysis and findings of this philosophy/autoethnography.

Creative Vignette

After class at the school garden, a veteran teacher puts her hand in soil, cool and yielding between her fingers. This soil is well hoed and watered. Life abounds below and above each flower. Below, bacteria, fungi, algae, and earthworms make the soil a living thing. Bees buzz above. This corner of the garden, with its orange mums, rosy autumn joys, purple turtleheads, and, soaring above them all, tall yellow sunflowers, has become her favorite. It is a sanctuary of sights, scents, and sounds. It is a sanctuary from the fast-paced, test-driven place of schooling inside. This garden recalls her to childhood, spending long afternoons with her grandmother gardening, watering plants, learning about soil and seeds, and pulling weeds. Gardens teach much about life. The garden was then, and continues to be, an education in living and loving on soil.[3] It teaches what it means to live well in a place, to balance culture and nature, and to become a sustainable part of a local ecology.

She wants her students to experience this. She hopes her pedagogy provides an opportunity for them to cultivate deep connections with soil. The bees, so crucial to each flower's growth, gently buzz around her. Pulling weeds by the roots and, emptying the school's rain barrel, watering flowers, she led her students in a discussion about bees' buzzing—how it's a natural result of flying; how buzzing becomes more aggressive when bees feel threatened; and that buzzing is important to flower pollination. The buzzing of bees is a song of life, guaranteeing new beauty each year. The touch of soil and the sounds of the garden teach the gardener how to live and love.

Historical Precedents

The work of music educator Satis Coleman (1878–1961) provides a seed for eco-literate music pedagogy. It was a seed that took root in the 1920s. A seed that currently lies dormant. She taught in private piano studios and music classrooms in Washington, DC and New York City, at Teachers College, Columbia University, and at the Lincoln Lab School. Her creative music method included field trips to understand the musics of others, instrument construction, and improvisation (Boston 1992). She published a number of books and articles, and her pedagogy was influential in the first half of the 20th century, but was lost to the annals of history. Perhaps her ecological vision for music education can be understood as our profession's road not taken (echoing Rachel Carson's quote). Some seeds sprout after remaining dormant for decades. If her ideas had remained influential, a new eco-literate music pedagogy might not be required in the first half of the 21st century. Music education might already be a sustainable, place-rooted, ecologically literate profession.

Music has an opportunity to be an important part of the education of children, because hearing is one of the five senses. This statement is not a validation of music in schools. Parents and communities educate children to be seeing, hearing, tasting, scenting, and feeling adults. Experienced arts such as crafts, foods, and musics often define a culture. Hearing is a primary way people experience the world, and music education may be able to help cultivate listening. Coleman's (1939) music pedagogy involved listening to nature *as* music.

> Many people go through life deaf to some of the most beautiful sounds in Nature. . . . All this rare music is missed, probably because the traveler was never taught, as a little child, to listen for these things and to love Nature with his[4] ears as well as with his other senses. (95)

Pedagogically, Coleman (1939) described activities that connect children to nature, including the "dramatization of songs . . . representing an animal or other character" (134–135), and promoted summer camps so children experience "out-of-door life, and all other healthy summer activities which are so fine for a child's health and social development" (129). She viewed nature's music as an opportunity for cultivating spiritual wellbeing, "a bit of understanding of the voice of God" (92).

There are many assumptions in Coleman's writing about music and nature that seem progressive, even today. In the praxial position, a position I begin with, music is understood as something people do (see Elliott and Silverman 2015; Goble 2010; Small 1998). To Coleman, music is more than something *people* do, but something all of nature does (humans, non-human animals, plants, and natural systems such as wind). Considered as *intentionality*, drawn from Husserl's (1983) phenomenology (*ego cogito cogitata*: a human intends [music]; this structure is extended in Chapter 5, "Spiritual Praxis"), Coleman's position is radically ecocentric, ecology-centered (humans intend music; but also, chickadees intend music; whales intend music; even the moon, wind, and waves intend music). This has implications for what music is (music as *something people do* becomes music as something people, non-human animals, plants, and natural processes do). And this ecocentric position has basic implications for what humans find sonically valuable. The praxial position demands we consider *other* musics "*on their own terms*, thereby gaining intercultural knowledge" (Goble 2010: 279, emphasis in original).

If considering music praxially[5] helps music teachers and learners value people outside of the Western world (because we value not only classical music, but also gamelan, mariachi, hip-hop, bluegrass, etc.), an ecocentric position insists we value musics beyond the anthropocentric. Extending Goble's position, and inspired by Coleman's ecocentric understanding of musics, we can study non-human musics *on their own terms*, thereby gaining ecological knowledge (see Chapter 4, "Deep Ecology").

Coleman's pedagogy seems in line with the work of Canadian composer R. Murray Schafer (1994), which began in the 1970s. Schafer's *soundscape* influenced the contemporary disciplines of ecomusicology and soundscape ecology (the scientific study of ecological systems through sound).[6] Schafer suggests that studying a society's soundscape can inform us about social conditions and trends, and that the world's soundscape is getting worse owing to noise pollution. That we have learned to ignore noises is problematic. He recommends a practice of clairaudience, "clean hearing" (11), to assuage noise pollution.

Schafer's (1994) discussion of the natural soundscape and the sounds of life are valuable to eco-literate music pedagogy. He categorizes the natural soundscape into the voices of the sea, the transformations of water, voices of wind, the miraculous land, unique tones, and apocalyptic sounds (15–28). For instance, when discussing voices of wind, Schafer writes, "The wind is an element that grasps the ears forcefully. The sensation is tactile as well as aural" (22). This connects to his pedagogy: "Sometimes I ask students to identify moving sounds in the soundscape. 'The wind,' say some. 'Trees,' say others. But without objects in its path, the wind betrays no apparent movement. It hovers in the ears, energetic but directionless" (23). Wind's contributions to the soundscape were some of the first sounds camp students noticed.

The sounds of life include birdsong, insects, the sounds of water creatures, and the sounds of animals in Schafer's (1994) categorization (29–41). Human speech, especially noticeable in onomatopoeia and music, can be studied as imitation of the sounds of life (e.g., Kīsêdjê music imitates non-human animal sounds; see Chapter 3, "Ecomusicology"). Speech is limited when it comes to imitating the sounds of life. Music is important because "it is in music alone that man finds the true harmony of the inner and outer world" (42). It seems music connects humans' internal, psychological, and spiritual lives with external soundscapes.

Ecology in Music Education

Schafer (1994) identified rural soundscapes as exhibiting better quality (43). Noise pollution degrades urban soundscapes, which negatively impacts people living in urban places. It is not surprising then that the work of Vincent C. Bates (2011, 2013a, 2013b, 2014, 2016) on rural ideals, such as living sustainability within a place (Bates 2013b), initiates music education's current move toward eco-literate music pedagogy. In this body of work, Bates connects ecological and social class, including in how social mobility means moving rural people to urban places (Bates 2011: 112). Bates expounds a hierarchy of musical engagements that places "school over outside-of-school, formal over informal, presentational over participatory, highbrow over lowbrow" (Bates 2011: 109), and recommends an agrarian perspective, which finds beauty in nature, and that values "households, neighborhoods, small towns" (Bates 2013a: 85) allowing students to live in "comfortable and original musical cultures" (Bates 2014: 321). This rural music education provides a model to begin thinking about music education *about*, *in*, and *for* the environment (Bates 2013a: 80), which provides an opening for eco-literate music pedagogy, which aims to be *placed*[7] and sustainable.

In line with the move toward eco-literate music pedagogy and because global ecological crises stem from inequalities in human and non-human "relationships," Charlene A. Morton (2012) recommends a music education philosophy for "all my relations" (472). Morton draws on Deweyan theory to suggest music education can embrace social and ecological justice, can identify its ecological impact, and can be a force for solving the world's problems. Morton's work leads to eco-literate music pedagogy's focus on ecological justice, and on recognizing the power of music as a way of knowing and solving the ecological crises.

Guiding Questions

In the 1990s, music educators began to conceive of music education from a praxial position[8] (e.g., Alperson 1991; Bowman 1998; Elliott 1997; Elliott and Silverman 2015; Regelski 1998). This turn can be seen as a leaving[9] and as a returning.

As a *leaving*, praxial music education called into question the values of music education as aesthetic education—values that seemed inextricably linked to classical music, intrinsic (to music) values, and music as an object (see Regelski 2016). Regelski suggests the aesthetic view is an ideology "so firmly taken for granted and engrained in many teacher's minds that any evidence or departure from its status quo catechisms is often rejected out of hand" (11). In contrast, praxialists view music as something people do,[10] emphasizing human, cultural, and social dimensions, and diversity (Alperson 1991; Bates 2013a; Elliott and Silverman 2015; Regelski 2016; Small 1998).

As a *returning*, praxialists often look to pragmatic philosophers, especially John Dewey,[11] prioritizing social aspects of music education (see Silverman 2012). Thus pragmatic music educators from the early 20th century like Satis Coleman, whose philosophies seem to have been replaced by aesthetic theorists, can be linked to current social valuing in music education. Praxial music education is essentially democratic.

But what is praxis in music education? Praxis is first and foremost a reflective action, "becoming a *reflective practitioner*" (Regelski 2016: 31). It begins with reflection as an ethical action. According to Elliott and Silverman (2015), praxis relates to musics in three ways: "(1) critically reflective and informed actions that are (2) embedded in and creatively responsive to both traditional and ever-changing musical/cultural/social values and (3) understood, taught, guided, and applied ethically and democratically for positive improvement of students' personal and musical-social-community lives" (17). Following this conception, in the current book I am critically reflecting on ecologically informed actions. This is what autoethnographic research is about—reflecting honestly on actual experiences. These actions are embedded in and creatively responsive to both tradition and ever-changing values. They take place in schooling settings, including a university-level elective on Western music, and a summer choir camp. Also, the praxial rationale of this autoethnography is *for* the positive improvement of students' personal and musical-social-community lives. It is the ecological crises that seem to characterize the 21st century as much as, if not more than, other social challenges facing humanity. As an expression of ecological consciousness and ecological literacy, the current book provides a unique contribution to praxial music education.

Enacting music pedagogy *for* ecological consciousness and ecological literacy is praxis—reflection and action for positive transformation of humanity in regard to the ecological crises we face. The guiding questions for this autoethnography are: What does it mean to embrace the impulse to teach music *for* ecological literacy? What is it like to theorize eco-literate music pedagogy? What is learned through enacting this pedagogy? How do the impulsion, the theorizing, and the enacting relate to one another? These questions help me to recognize the urgency of teaching music for ecological literacy as humanity faces its greatest existential crises since the invention of nuclear weapons. They are "answered" in this book through reflections on my pedagogy in the first year of enacting eco-literate music pedagogy, constructively through the creative vignettes imagining different ways this pedagogy might be, artistically through the poems that end each chapter, and theoretically through scholarly literature (in all chapters, but heavily in Chapters 2 and 3, which function a bit as literature reviews) and building logical arguments (especially in Chapters 1, 4, and 5) for a spiritual praxis of music education on soil.

Gardening and Wilderness

As metaphors, *gardening* can be used to clarify *ecological consciousness*, and *wilderness* can be used to clarify *ecological literacy*. To understand eco-literate music pedagogy, understanding how I conceive of these two ideas becomes important. At a fundamental level, the idea of consciousness is awareness. People become aware of objects or ideas through five senses—sight, hearing, touch, taste, and smell.[12] Because of the small scale of consciousness—the individual person is the unit of measurement—consciousness can be seen through the lens of *place*: a person in a place becomes conscious of objects and ideas experienced through his/her five senses.

If two metaphors guiding ecological discourse are *wilderness* (such as the work of Arne Naess) and *gardening* (such as the work of Wendell Berry and Michael Pollan), consciousness is understood through the act of gardening. Gardening is experienced through the five senses (including hearing). We choose which nature thrives in our gardens; we plant seeds and pull weeds. We're guided by the wisdom of people who have lived on this soil before, and who live nearby. Ecological consciousness is a process of living well in a place, of balancing our cultural (including aesthetic) desires with nature near us. We choose what beauty to cultivate in our gardens. Ecological consciousness can and should involve forming positive experiences through education. This metaphorical use employs a simplified conception of gardening. The metaphor transforms by considering specific schools of gardening such as organic, raised-bed, no-till, or permaculture. However, all gardening techniques say something about how people live with nature. To control weeds one gardener sprays pesticide and another grows cover crops. To aerate soil one gardener plows and another plants radishes. These choices say something about how each person understands his/her relationship to nature. Each is specific and experienced. Ecological consciousness is experienced as cultural.

In contrast, ecological literacy involves abstraction. In this way it is more like *wilderness* than *gardening*. Wilderness is necessary, but not lived in by us (human life makes a place un-wilderness or at least less-wilderness). Questions of wilderness invoke questions of policy. For instance, the U.S. Bureau of Land Management identifies wilderness as "special places where the earth and its community of life are essentially allowed to function without manipulation. They retain their primeval character, without permanent improvements and offer outstanding opportunities for solitude or primitive and unconfined recreation."[13] As policy, wilderness contains paradoxes. How does a place lacking human manipulation endure as recreation? Once we talk about human use of wilderness, we can ask, "Is there any wilderness left?" *Wilderness* is theoretical, which leads to specific policies (e.g., the number of people who can enter; the permissibility of hiking or horses), which can be modified.

In general, literacy means coming to understand many aspects of society (and not merely reading and writing). Many of the ecological crises—such as climate change, the Great Pacific Garbage Patch, species loss—do not necessarily involve individual consciousness of the ecological crises. These global crises may involve individual experiences of manifestations of a crisis. When a Yup-ik Eskimo loses his/her home to melting tundra, or when an Arizona farmer runs short of water, climate change is manifested, but climate change as a concept requires an ecologically literate citizen who connects these experiences to a theory. An ecologically literate citizen, especially in a suburb, may not actually experience any of these manifestations (e.g., waste can be artfully kept out of sight). Thinking historically and re-theorizing are important. When manifestations occur differently than previously theorized, the ecologically literate citizen understands a newer, more accurate theory (in recent history, the greenhouse effect was replaced by global warming and then climate change, each providing a better theory). Ecological literacy as praxis involves looking courageously at the ecological crises (reflection) and identifying ways to alleviate these crises (action). Both ecological consciousness/gardening and ecological literacy/wilderness are essential components of eco-literate music pedagogy, because people crave both experiences and sense making.

Essential Structure of Eco-Literate Music Pedagogy

Previously I theorized eco-literate music pedagogy (Shevock 2015b: 17):

1 connecting to local places;
2 experiencing music and nature in connected, meaningful, and ethical ways;
3 developing ecological consciousness by ritualizing and creating music rooted in soil;
4 connecting to the planet more broadly by connecting local understandings to global ecological crises.

Eco-literate music pedagogy begins with helping students become conscious of local places. This is no small task for people focused on computer and cellphone screens. Because music is a way humans come to understand the world around them through sound and other senses, the sonic experiences of ecological consciousness are a fundamental aspect of eco-literate music pedagogy. Drawing on Coleman and Schafer, music teachers and students study the musics that surround them. This leads to teachers and students living well in place, connected to the health of local environments and dedicated

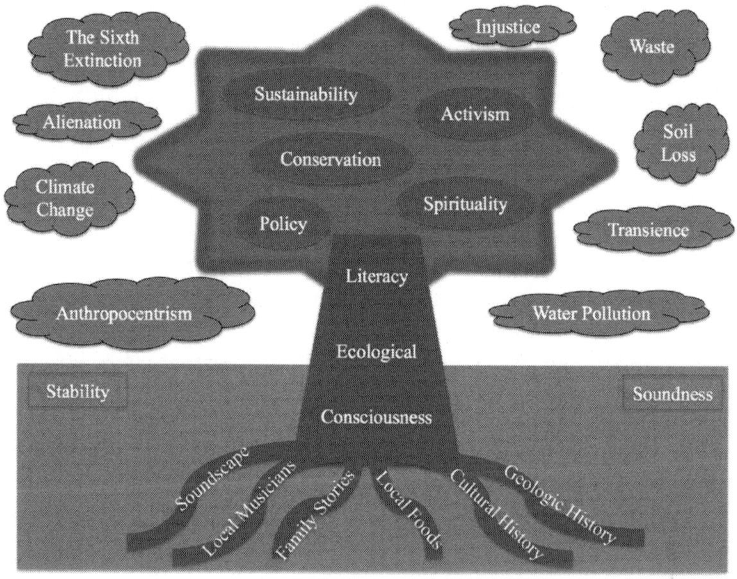

Figure P.1 Essential Structure of Eco-Literate Music Pedagogy

to protecting those places. Music teachers and students can be inspired by the musics of non-human life to create musics for performance. These can provide an opportunity for pleasure and a more scientific understanding of these places (what non-human animals music in these places), and then can be connected to broader theories (such as climate change, waste, soil use, transportation, suburbanization), the understanding of which typifies an ecologically literate citizen.

Reflecting with Purpose

The purpose of this self-study is to explore the theory underlying eco-literate music pedagogy in juxtaposition with personal experiences of the author, a music teacher enacting this pedagogy. According to ecotheologian Thomas Berry (1999), ecological devastation is primarily caused by "a mode of consciousness that established a radical discontinuity between the human and other modes of being and the bestowal of all rights on humans" (4). To understand my consciousness, this book utilizes reflective self-study as I try to expand my conception of modes of being and of rights. This philosophy/autoethnography represents my evolution as a music teacher and, in turn, modifies theory with practice.

Reflective Writing in Environmentalism

Reflective writing dominates the environmental literature. Thoreau provided the model for much of this writing. Books such as *Walden* (Thoreau 1995) and essays such as "Walking" (Thoreau 1851) characterize self-reflective writing that moves theory. Thoreau's writing pushes Emerson's transcendental philosophy toward the biocentric, placing humankind *in* nature, rather than above it. Titon (2015) discussed Thoreau's movement away from transcendentalism and toward a more scientific and experiential philosophy: "his thinking gradually moved away from spiritual correspondences and toward scientific truths experienced as patterns in nature. Unusually oriented to sensory experience, listening became an important means toward this knowledge" (144). In *Walden*, Thoreau wrote a chapter on "Sounds," which can be seen as a basis for sound studies, soundscape ecology, ecomusicology, and eco-literate music pedagogy.

Notable reflective writing includes the work of Sierra Club founder John Muir (1838–1914) and environmental advocate Edward Abbey (1927–1989).[14] John Muir's journals provide insight into the mind of the conservationist's camping trips with President Theodore Roosevelt, the discovery of Glacier Bay, and his insight into American philosopher Emerson (Teale 2001). Abbey (1991)—like Pollan (2003) a decade later—begins his reflective insights with a critique of Thoreau, critiquing his mystical incoherence and his timidity. But ultimately, Thoreau needs to be approached: "I look for his name in the water, his face in the airy foam. He must be here. Wherever there are deer and hawks, wherever there is liberty and danger, wherever there is wilderness, wherever there is a living river, Henry Thoreau will find his eternal home" (48). This type of writing seems praxial, understanding theory through actual experience.

More recent reflective writing has considered such topics as gardening and farming (Berry 2010; Kingsolver 2007; Pollan 2003), beekeeping (McKibben 2013), and living without money (Sundeen 2012). Even when no explicit theory is drawn, this body of writing puts forth a variety of theories. Many writers use Wendell Berry's essays to develop agrarian theory, including in music education (e.g., Bates 2013a, 2013b, 2016). The message underlying Sundeen's (2012) narrative is that living simply can lead to living better. Kingsolver (2007) promotes the importance of locavorism, living and growing food locally. McKibben (2013) tells his story during a time when he was shifting roles from environmental writer to activist, while at the same time purchasing a plot of land and learning about beekeeping. The message centers on the importance of global action, acting to change government policy, while connecting to a local place, concurring with the current concepts of wilderness and gardening.

Autoethnographies in Music Education

The first autoethnography I read was Kruse's (2012) outlining his attempts to learn mandolin online. Caught up in this story of using only online resources, an act that deeply challenges institutional music education, he described not only successes, but also frustrations and challenges. Autoethnography can be about sharing vulnerability (Bartleet and Ellis 2009). Kruse's research set me on a course to question the possibilities, roles, and means of scholarship in music education. My own autoethnographic research was, following Kruse's model, an analysis of my praxis (Shevock 2015a). Later I began to realize the more evocative possibilities for autoethnography (such as proposed by Bartleet and Ellis 2009; Denzin 2014; Ellis 2004; Gouzouasis et al. 2014), though I have strived for a balance between analysis and expressiveness (Shevock 2016), which leads me to philosophy/autoethnography.

Music education has seen a sharp increase in autoethnography. These studies have been published in many peer-reviewed publications, including *Action, Criticism, and Theory for Music Education* (Bates 2011; Lamb 2014; Shevock 2015a, 2016); *British Journal of Music Education* (Bartleet and Hultgren 2008; Mapana 2011); *Bulletin of the Council for Research in Music Education* (Taylor 2014); *International Journal of Education and the Arts* (Hendricks 2013; Gouzouasis et al. 2014); *Journal of Music, Technology and Education* (Kruse 2012); *Journal of Research in Music Education* (Nichols 2016); *Music Education Research* (Thompson 2015; Webb 2008); *Qualitative Report* (Nethsinghe 2012; Sutherland 2015); and *Research Studies in Music Education* (Heuser 2015; Saldaña 2008).

Music education autoethnographies have been used to passionately explore issues of gender (Lamb 2014; Nichols 2016; Taylor 2014), race (Shevock 2016; Thompson 2015), class (Bates 2011; Shevock 2016), and emotional pain (Saldaña 2008; Shevock 2016; Taylor 2014); to analyze storytelling itself (Gouzouasis et al. 2014; Nichols 2016); to interpretively explore creative pedagogy (Bartleet and Hultgren 2008; Heuser 2015; Kruse 2012; Shevock 2015a; Sutherland 2015); to expand the cultural dimensions of music education (Mapana 2011; Webb 2008); and to challenge current teaching practices (Hendricks 2013; Nethsinghe 2012; Shevock 2016). These autoethnographies accomplished many purposes and, importantly to this book, describe and challenge theory (Bates 2011; Lamb 2014; Shevock 2015a, 2016). Bates's work uses rural theory, Lamb's feminism, and Shevock's Freirean and Illichan theory.

Autoethnography can be seen as part of a new ethos in music education research, "ethical, spiritual, and heartful . . . uncontrolled, rebellious, and gloriously perfect imperfection" (Gouzouasis et al. 2014: 18). Autoethnographies can be, in contrast, analytic and focused on developing and challenging social theory (Anderson 2006). In the current book, I hope to balance some of the more analytic elements of autoethnography (such as exploring the possibilities of a new pedagogical model) with some of the more *ethical* (the ecological crises are deeply ethical concerns), *spiritual* (I recommend eco-literate music pedagogy as a spiritual praxis), and *heartful* (developing a new philosophy in practice is an emotional endeavor to which I lend my full *self*, including my professional reputation). I call this balancing act—of rugged interpretation, pliable heartfulness, and theory building—philosophy/autoethnography.

Philosophy/Autoethnography

As an autobiographical research method, autoethnography is typified by its use of insider knowledge, nuance, and accessible prose (Ellis and Adams 2014). Autoethnography has emerged as an important qualitative research methodology, compatible with our current concern with identity politics (259). Autoethnography aims for "triadic balance" (Chang 2008: 48) between self (*auto-*), culture (*-ethno-*), and research method (*-graphy*). Accordingly, "autoethnographers hope to gain cultural understanding of self and others directly and indirectly connected to self" (49). Music teaching institutions provide one *culture* in this analysis, but other cultures such as family and non-human species cultures emerged as essential. In the autoethnographic tradition Ellis (2004) promotes, the author writes in the first person, becoming a storyteller (30). This is consistent with how I understand *good* writing, which doesn't hide the actor (the reader can easily identify the sentence structure to identify who did what), and in the current book I have chosen to utilize the "ethnographic I" extensively.

Structured as memoirs, autoethnographies utilize the conventions of storytelling. Denzin (2014) identifies these conventions: "People depicted as characters; a scene, place, or context where the story occurs; an epiphany or crisis that provides dramatic tension . . . a temporal order of events; [and] a point or moral to the story" (4). As an example of how these conventions have guided this book, my "epiphany" occurred after completing my dissertation, when I longed to find coherence between my experiences with nature, music, and pedagogy. However, because this philosophy/autoethnography represents a combination of philosophy and autoethnography, the memoir aspects are divided with more theoretical writing. While a memoir may allude to philosophical truths, an aim of philosophy is to make those explicit.

In my research—the present study is my third using autoethnography—I concluded, "because of autoethnography's subjective nature, it is well suited for exploring theoretical concepts that might be difficult to quantify" (Shevock 2016: 47). Eco-literate music pedagogy is rooted in this body of theory-driven self-reflective research, which I am labeling philosophy/autoethnography. Philosophy/autoethnography is a subgenre of autoethnography that embraces the evocative nature of autoethnography and makes explicit and builds theory.

Data analyzed for this philosophy/autoethnography were collected from February 2015 through September 2016 (1 year, 7 months) and include reflections kept in a journal of my teaching, and student assignments. Teaching experiences include two youth summer choir camps, and two semesters (four sections, teaching approximately thirty students in each) of teaching MUS 005: Introduction to Western Music at Penn State Altoona. The reflections were done as soon after teaching as possible, to help ensure accuracy. The choir camp listened to nature at a park and co-composed music with voices and natural found sounds.[15] The Introduction to Western Music assignments included a written listening response to Hunter Hensley's *Requiem for the Mountains*,[16] which features Gregorian chant with images of the ecological damage to Kentucky mountains due to mountaintop removal mining; a reflection comparing Hensley's requiem with Mike Townsend's *They Call It Progress*,[17] a folk song that deals with the same topic; and a sound ecology reflection on the "sounds of nature" as if they were an intended musical composition (using music elements, such as rhythm, pitch, and melody, and identifying "instruments" used, such as ducks, wind, and construction equipment), and various additional listening and discussions around nature and music. These data were reconstructed into various stories, which became the base unit of analysis for writing the memoir aspect of this philosophy/autoethnography.

Data analysis is presented as reflections (reconstructing actual experiences) and creative vignettes (constructing fictional experiences to elucidate a *truth* or *future possibility* for eco-literate music pedagogy). Each chapter is ended with a poem, an artistic representation of analysis, following the lead of many autoethnographies (such as de Vries 2007; Shevock 2015a, 2016). These analyses are divided with theoretical writing—insight drawn from other scholars and arguments I construct.

In Chapter 1, "Philosophy on Soil," Ivan Illich, place-based education, and a theory of the commons provide a pedagogical foundation for ethical action in place. A music education on soil is advanced. Eco-literate music pedagogy is framed as interdisciplinary in Chapter 2, "Ecological Literacy." In Chapter 3, "Ecomusicology," ecomusicology, a subfield of both musicology and ethnomusicology, is described. In ecomusicology, the environment is portrayed as essential to understanding music: environments inspire music compositions and environmentalists use musics in their activism. Expanding the concept of music education on soil, Chapter 4, "Deep Ecology," draws from the deep ecology platform to frame an ecocentric valuing of non-human musicking. Music education on soil as a deep ecology leads to the premise of a praxial spirituality of eco-literate music pedagogy in Chapter 5, "Spiritual Praxis." Living the good life involves an expanded understanding of self: living for all our kin—human, non-human, and place.

Mysteries Pondered in a Garden

agony, the garden
work cut from life. cynicism's burden. walking, listening, wildlife
  valuing on soil

scourging, the pillar
aching of rebirth. cocooned caterpillar. vigor and earth
  disentangling weeds

crowning with thorns
self-taunting thoughts. excesses are shorn. a seed, a fresh plot
  harking birdsongs

carrying the cross
setting pen to paper. a plot becomes acres
  pondering ground

crucifixion and death
completion and joy. ecological breath. detested, enjoyed
  fulfilled and organic

Notes

1 From Carson's (2002) foundational environmental science activism book *Silent Spring* (277).
2 Koza (2006) calls for a process of "sifting and winnowing" (36) to curb consumerism in music education. This seems consistent with eco-literate music pedagogy.
3 Ivan Illich (with Sigmar Groeneveld, Lee Hoinacki, and other friends) (1990) suggests "our generation has lost its grounding in both soil and virtue" (1). In the recommended *philosophy on soil*, discussed in Chapter 1, "Philosophy on Soil," actions should be conscious of local traditions and place-bounded. A school garden can become, for teachers and students alike, an opportunity to cultivate such virtue, while many school activities seem to promote non-soil-bound actions, and are ecologically unsustainable.
4 Here I want to note that, while much of the literature used in this book employs gender-specific language, I prefer gender-inclusive language. However, I have made the choice to avoid including *sic* after each referenced occurrence, to improve flow for the reader. This was a hard choice for me, but was made as a result of standard practice in historical writing.
5 From some perspectives, praxial philosophies can be understood as anthropocentric. This is explored a little in this book. A full analysis of whether or not these are anthropocentric is beyond the scope of this book. This book aims to construct a philosophy of music education on soil, not critique praxial philosophy. Subjectively, praxial music education philosophy has inspired me to begin considering eco-literate music pedagogy. I understand a philosophy of music education on soil as praxial, which will be illuminated in Chapter 5, "Spiritual Praxis."
6 For instance, studying two common birds in the Eastern U.S., the wood thrush and red-eyed vireo, might require soundscape ecology because both birds are rarely observable—they rarely leave dense undergrowth—but are prominent singers contributing to the soundscape.
7 This autoethnographic approach to understanding eco-literate music pedagogy tries to echo a tradition of writing in music education that is placed not only in physical space, but in the sense of local validity: "'locally' valid and persuasive without claiming applicability to all musical situations for all time" (Bowman 1998: 16).
8 Goble (2010) constructed a more robust description of this praxial movement that focuses more specifically on Elliott's and Regelski's influences and distinctions that might be of interest to readers (237–245).
9 For instance, Elliott (1997) identified his praxial philosophy of music education as "an alternative" (1) to Bennett Reimer's music education as aesthetic education. Elliott aimed to stimulate critical thinking: "There are many ways to conceptualize the nature and significance of music education. There is ample room for alternative views" (3). Regelski (1998) referred to praxial philosophy as "a critical challenge to the main traditions of aesthetic theory and their implications for music education" (22).
10 Aesthetic considerations, from the praxial perspective, are subservient to social considerations. For instance, Alperson (1991) wrote,

> The basic aim of a praxial philosophy of music is to understand, from a philosophical point of view, just what music has meant to people, an endeavor that includes but is not limited to a consideration of the function of music in aesthetic contexts (234).

11 In addition to Dewey, Goble (2010) draws from Dewey's teacher C. S. Peirce.
12 These are the traditionally recognized senses. There seem to be more than five senses, including thermoception, proprioception, nociception, equilibrioception, and mechanoreception (link: https://en.wikipedia.org/wiki/Sense). Perhaps musicking is experienced using all of these senses.
13 Link: https://www.blm.gov/programs/national-conservation-lands/Wilderness-Areas.
14 Contrasting reflective writing, Rachel Carson's *Silent Spring*, so important to the modern environmental movement, is expressly not reflective. It may be that, since she was dying of cancer while writing, she hoped to avoid the DDT industry claiming she was merely expressing a vendetta. *Silent Spring*, while artfully written, portrays its message as objective.
15 The students' compositions can be found online at YouTube (Year 1 link: https://youtu.be/dNx5zZrIo9s; Year 2 link: https://youtu.be/5N4txoPzVUo). They performed these at their end-of-camp concert.
16 Link: https://youtu.be/g8gJrTe2O18.
17 Link: https://youtu.be/50v-nckcoSg.

References

Abbey, Edward. 1991. *Down the River*. New York: Plume.

Alperson, Philip. 1991. "What Should One Expect From a Philosophy of Music Education?" *Journal of Aesthetic Education* 25, no. 3 (Autumn): 215–242.

Anderson, Leon. 2006. "Analytic Autoethnography." *Journal of Contemporary Ethnography* 35, no. 4 (August): 373–395. DOI: 10.1177/0891241605280449.

Bari, Judi. 1994. *Timber Wars*. Monroe, ME: Common Courage Press.

Bartleet, Brydie-Leigh, and Carolyn Ellis, eds. 2009. *Music Autoethnographies: Making Autoethnography Sing/Making Music Personal*. Bowen Hills: Australian Academic Press.

Bartleet, Brydie-Leigh, and Ralph Hultgren. 2008. "Sharing the Podium: Exploring the Process of Peer Learning in Professional Conducting." *British Journal of Music Education* 25, no. 2: 193–206. DOI: 10.1017/S0265051708007924.

Bates, Vincent C. 2011. "Sustainable School Music for Poor, White, Rural Students." *Action, Criticism, and Theory for Music Education* 10, no. 2: 100–127.

——. 2013a. "Music Education Unplugged." *Action, Criticism, and Theory for Music Education* 12, no. 2 (September): 75–90.

——. 2013b. "Drawing from Rural Ideals for Sustainable School Music." *Action, Criticism, and Theory for Music Education* 12, no. 1 (April): 24–46.

——. 2014. "Rethinking Cosmopolitanism in Music Education." *Action, Criticism, and Theory for Music Education* 13, no. 1 (March): 310–327.

——. 2016. "Toward a Sociology of Music Curriculum Integration." *Action, Criticism, and Theory for Music Education* 15, no. 3 (May): 1–14.

Berry, Thomas. 1999. *The Great Work: Our Way into the Future*. New York: Bell Tower.

Berry, Wendell. 2010. *What Are People For?* Berkeley, CA: Counterpoint.

Boston, Sheila C. 1992. "Satis N. Coleman (1878–1961: Her Career in Music Education." PhD dissertation, University of Maryland, College Park. 9315605.

Bowman, Wayne. 1998. "Universals, Relativism, and Music Education." *Bulletin of the Council for Research in Music Education* 135 (Winter): 1–20.

Carson, Rachel. 2002. *Silent Spring*. Boston, MA: Mariner Books. [Originally published in 1962.]

Chang, Heewon. 2008. *Autoethnography as Method*. Walnut Creek, CA: Left Coast Press.

Coleman, Satis N. 1939. *Your Child's Music*. New York: John Day Company.

Denzin, Norman K. 2014. *Interpretive Autoethnography*. Los Angeles, CA: Sage.

de Vries, Peter. 2007. "Using Poetry as Data Representation to Explore Music Opportunities Parents Provide for Their Children at Home." *Music Education Research* 9, no. 1 (March): 17–34.

Elliott, David J. 1997. "Continuing Matters: Myths, Realities, Rejoinders." *Bulletin of the Council for Research in Music Education* 132 (Spring): 1–37.

Elliott, David J., and Marissa Silverman. 2015. *Music Matters: A New Philosophy for Music Education*, 2nd edition. New York: Oxford University Press.

Ellis, Carolyn. 2004. *The Ethnographic I: A Methodological Novel about Autoethnography*. Walnut Creek, CA: Altamira Press.

Ellis, Carolyn, and Tony E. Adams. 2014. "The Purposes, Practices, and Principles of Autoethnographic Research." In *The Oxford Handbook of Qualitative Research*, edited by Patricia Leavy, 254–276. New York: Oxford University Press.

Goble, J. Scott. 2010. *What's So Important about Music Education?* New York: Routledge.

Gouzouasis, Peter, Danny Bakan, Jee Yoen Ryu, Helen Ballam, David Murphy, Diana Ihnatovych, Zoltan Virag, and Matthew Yanko. 2014. "Where Do Teachers and Learners Stand in Music Education Research? A Multi-Voiced Call for a New Ethos of Music Education Research." *International Journal of Education and the Arts* 15, no. 15 (October): 1–23.

Hendricks, Karin S. 2013. "Songs My Student Taught Me: Narrative of an Early Childhood Cello Teacher." *International Journal of Education and the Arts* 14, no. 1.4.

Heuser, Frank. 2015. "Pipe Dreams, Ideals and Transformation in Music Education: Lessons from the Field." *Research Studies in Music Education* 37, no. 2: 215–231. DOI: 10.1177/1321103X15614322.

Husserl, Edmund. 1983. *Ideas Pertaining to a Pure Phenomenology and to a Phenomenological Philosophy*, First Book, translated by F. Kersten. Dordrecht: Kluwer Academic Publishers.

Illich, Ivan. 1990. "Declaration on Soil: A Joint Statement, Drafted in Hebenshausen, Germany, December 6, 1990, in Collaboration with Sigmar Groeneveld, Lee Hoinacki and Other Friends." Online: http://davidtinapple.com/illich/1990_declaraion_soil.PDF.

Kingsolver, Barbara. 2007. *Animal, Vegetable, Miracle: A Year of Food Life*. New York: Harper Perennial.

Koza, Julia Eklund. 2006. "'Save the Music'? Toward Culturally Relevant, Joyful, and Sustainable School Music." *Philosophy of Music Education Review* 14, no. 1 (Spring): 23–38.

Kruse, Nathan B. 2012. "Locating 'The Road to Lisdoonvarna' via Autoethnography: Pathways, Barriers and Detours in Self-Directed Online Music Learning." *Journal of Music, Technology and Education* 5, no. 3: 293–308. DOI: 10.1386/jmte.5.3.293_1.

Lamb, Roberta. 2014. "Where Are the Women? And Other Questions, Asked Within an Historical Analysis of Sociology of Music Education Research Publications: Being a Self-Reflective Ethnographic Path." *Action, Criticism, and Theory for Music Education* 13, no. 1 (March): 188–222.

Mapana, Kedmon. 2011. "The Musical Enculturation and Education of Wagogo Children." *British Journal of Music Education* 28, no. 3 (November): 339–351. DOI: 10.1017/S0265051711000234.

McKibben, Bill. 2013. *Oil and Honey: The Education of an Unlikely Activist*. New York: Times Books.

Morton, Charlene A. 2012. "Music Education for 'All My Relations.'" In *The Oxford Handbook of Philosophy in Music Education*, edited by Wayne D. Bowman and Ana Lucía Frega, 472–491. New York: Oxford University Press.

Nethsinghe, Rohan. 2012. "The Influence of Informal Music Education in Teacher Formation: An Autoethnography." *Qualitative Report* 17, no. 98: 1–16.

Nichols, Jeananne. 2016. "Sharing the State: Ethical Dimensions of Narrative Inquiry in Music Education." *Journal of Research in Music Education* 63, no. 4: 439–454. DOI: 10.1177/0022429415617745.

Pollan, Michael. 2003. *Second Nature: A Gardener's Education*. New York: Grove Press.

Regelski, Thomas A. 1998. "The Aristotelian Bases of Praxis for Music and Music Education as Praxis." *Philosophy of Music Education Review* 6, no. 1 (Spring): 22–59.

——. 2016. "Music, Music Education, and Institutional Ideology: A Praxial Philosophy of Music Sociality." *Action, Criticism, and Theory for Music Education* 15, no. 2 (March): 10–45.

Saldaña, Johnny. 2008. "Second Chair: An Autoethnodrama." *Research Studies in Music Education* 30, no. 2: 177–191. DOI: 10.1177/1321103X08097506.

Schafer, R. Murray. 1994. *The Soundscape: Our Sonic Environment and the Tuning of the World*. Rochester, VT: Destiny Books. [Originally published in 1977.]

Shevock, Daniel J. 2015a. "Reflections on Freirean Pedagogy in a Jazz Combo Lab." *Action, Criticism, and Theory for Music Education* 14, no. 2 (August): 85–121.

——. 2015b. "The Possibility of Eco-Literate Music Pedagogy." *TOPICS for Music Education Praxis* 2015, no. 1: 1–23.

——. 2016. "Music Educated and Uprooted: My Story of Rurality, Whiteness, Musicing, and Teaching." *Action, Criticism, and Theory for Music Education* 15, no. 4 (August): 30–55. DOI: 10.22176/act15.4.30.

Silverman, Marissa. 2012. "John Dewey and James Mursell: An Introduction." *Visions of Research in Music Education* 21: 1–19.

Small, Christopher. 1998. *Musicking: The Meanings of Performing and Listening*. Middletown, CT: Wesleyan University Press.

Sundeen, Mark. 2012. *The Man Who Quit Money*. New York: Riverhead Books.

Sutherland, Andrew. 2015. "From Isolation to Collaboration: An Autoethnographic Account." *Qualitative Report* 20, no. 10: 1631–1645.

Taylor, Donald. 2014. "Learning to Sing Again: A Gay Widower's Journey." *Bulletin of the Council for Research in Music Education* 199 (Winter): 69–86.

Teale, Edwin Way, ed. 2001. *The Wilderness World of John Muir: A Selection from His Collected Work.* Boston, MA: Mariner Books.

Thompson, Jason D. 2015. "Towards Cultural Responsiveness in Music Instruction with Black Detained Youth: An Analytic Autoethnography." *Music Education Research* 17, no. 4: 421–436.

Thoreau, Henry David. 1851. "Walking." [Kindle DX version.]

——. 1995. *Walden, and On the Duty of Civil Disobedience*. Project Gutenberg. Online: http://www.gutenberg.org/ebooks/205. [Originally published in 1854.]

Titon, Jeff Todd. 2015. "Thoreau's Ear." *Sound Studies* 1, no. 1: 144–154. DOI: 10.1080/20551940.2015.1079973.

Webb, Michael. 2008. "Giles Apap's Mozart Cadenza and Expanding Musical Competencies of Twenty-First-Century Musicians and Music Educators." *Music Education Research* 10, no. 1 (March): 15–39. DOI: 10.1080/14613800701871348.

1 Philosophy on Soil

> We stand on soil, not on earth. From soil we come, and to soil we bequeath our excrements and remains. And soil—its cultivation and our bondage to it—is remarkably absent from those things clarified by philosophy in our western tradition. As philosophers, we search below our feet because our generation has lost its grounding in both soil and virtue.
>
> (Ivan Illich, Sigmar Groeneveld, Lee Hoinacki, and other friends, 1990[1])

Imagining Eco-Literate Music Pedagogy

Autumn 2003

A place of paradox, Riverview Park yields a seeming wilderness within the city. It's about a mile from my school. Partially fallen yellow maple leaves rustle below my feet. I take time to walk here after school. Slowly, softly, and tranquilly I walk. Walking becomes an *aim*, not a way to get elsewhere. Following a walking meditation described by Thich Nhat Hanh,[2] a popular Buddhist author, I imagine lotus flowers bursting forth at my feet. Birds, mammals, and insects fill the soundscape, and trees muffle the sounds of nearby highways. Up the hill, the Allegheny Observatory embodies science's long gaze, advancing knowledge to the deepest reaches of mechanically mediated vision. Down the hill, ruins of a zoo rest, rusty relics of a bygone day. I walk in this simulated wilderness, a paradox of future and past. Today is a day for realization.

On this walking meditation, my mind wanders. As I reflect on *wilderness*, birdsongs accompany me. I catch sight of a small predator: a hawk. A herd of deer in the brush, dozens large, has become unafraid of me. Their footsteps are nearly soundless. The noisiest inhabitant of the park is a squirrel, who, heedless of the stalking hawks, lives life fearlessly. There are lessons I learn here to living the good life. I open myself to receive those lessons. I listen.

Walking. Imagining. Suddenly, or maybe gradually, I realize this is not wilderness. Am I fooling myself? Perhaps it is the hawk and overabundance of deer drawing my attention to how safe I feel. No large predators, crucial for a self-sustaining wilderness, reside here. The old remains were a reminder that a century previous this park was a meadow, and before that, a farm. And these trees towering above me aren't indigenous inhabitants, but are younger, planted here to provide urban citizens with a wilderness simulation.

Spiritually, I benefit from losing myself in the simulacrum. After a stressful workday with too many hours sitting at a piano or desk teaching and grading, I walk and let simulated wilderness wash my tensions away. It's comforting to imagine ancient Iroquois meeting, hunting, and laughing together for generations under these same picturesque leaves. But it is an untruth. The shade provided by the Norway maples is not the shade those Iroquois knew. Indigenous trees were cleared for farming, for sustenance. Sadly, so were most of the Iroquois. My daily meditations would be less relaxed if wild wolves and mountain lions were not eradicated. Long after the land was used for sustenance, it rested, generally uncultivated (hence the abundance of invasive species).

Perhaps it is best to realize city parks have more in common with gardens than wilderness. Walking here helps me realize wilderness areas are important. And perhaps I shouldn't be walking on actual wilderness. There should be places where humans don't walk. I wonder how much *real* wilderness, humanless expanses, remain in our planet? How much should there be? These are political questions. Lost in speculation, I was cultivating ecological consciousness. Walking provided me a conscious experience on soil. Ecological literacy begins with cultivating my *self*, but it cannot end there. I should do more than walk to find answers to my questions about wilderness. I should read, talk, listen, change.

Autumn 1983

A mother stands with her children. The casket, surrounded by flowers, rests near a wall. Crying; her father, a lifelong carpet layer, had recently passed. The family returned from Maryland to rural Pennsylvania for the funeral. "Dad was involved in building both the Patton Park and Prince Gallitzin State Park. . . ."

The young boy listens silently, closely. A lesson in death and life takes root. The parks, trees, and wildlife grandpa protected can stand as monuments.

Even if only in my mind, I will appreciate them.

Spring 2015

Walking in Tudek Park with my young son, we echo birds. As a stay-at-home dad, I wrote a dissertation with my son sitting on my lap. He's becoming more active, more independent. What do I want to pass on to my son? What's important? What is virtuous? Standing on soil, I search below my feet. There, I see, hear, and feel life.

Philosophy on Soil

Historian and philosopher Ivan Illich problematized the institution of (Christian) values in Western society. He suggests, "the institutionalization of values leads inevitably to physical pollution, social polarization, and psychological impotence: three dimensions in the process of global degradation and modernized misery" (Illich 1970: 1). Schooling is one example. People have learned (reading, math, music) throughout history without institutions. Speaking generally, learning is good. As something we value, it can be called a value. Schools confound the value with institutions. Education commodifies learning by institutionalizing it. For Illich, environmental damage is the expected result. In contrast, a philosophy on soil is centered on traditions and place. Actions are virtuous to the extent that choices are made in practices "mutually recognized as being good within a shared local culture that enhances the memories of a place" (Illich 1990: 1).

Music education is not immune to the institutionalization of values. Professional educators teach students to presume music education, developing students into *homo educandus musicae*, "s/he who believes music education is a basic human need (everybody ought to be music educated) and that music education is a prerequisite to meaningful musicing" (Shevock 2016: 44). Reformers often include more diverse and popular musical genres, but seldom challenge institutionalization itself. Illich (2012) contends education is a *sacred cow*: "They all want more education, directed toward the sector which they represent" (128), and "We fight now, in the name of education, against a teaching profession which unwillingly constitutes an economic interest" (129). For musics, when a new genre is included in schools, people redefine themselves as *non-musicians* in regard to that genre. Schools (and not families or communities, or another naturally occurring organization of humans) label some people experts, and everybody else not. It seems like hubris to believe schooling institutions can take the place of living musical cultures in educating the next generation. Schooling happened with classical music, and more recently with jazz. At the current rate of institutionalization, we can expect future generations of hip-hop and bluegrass musicians to *need* diplomas. And hubris will pile upon hubris as these musicians become the institutionalized experts. Providing more music education, delivered by *unwilling* propagators of economic interests, is probably counterproductive if we want more musicking in communities.

Perhaps we need to "ask which society, which education, which form of religion, is beneficial for all life on the planet as a whole" (Arne Naess in Devall and Sessions 1985: 74). A study of small-scale, local musics (music learned somewhere besides music education institutions) might show a way forward for music education. That way forward will likely be rooted in humility, recognizing virtuous limits for our profession.

"An understanding of indigenous ways of seeing encourages rethinking the purposes of education" (Shirley Steinberg and Joe Kincheloe in Prakash and Esteva 2008: xvi). Abrokwaa (1999) shares three African environmental songs that honor Mother Earth, and suggests that environmental and social factors guide indigenous African musics. Matsunobu (2009) notes that Japanese *shakuhachi* flute making "is interdependent on the natural resources available in each place and cannot occur without a sustainable relationship with the land" (vi). To re-root music education in soil, teachers might humbly study the songs and teaching practices of *the other* to understand how indigenous people come to know their environments and live sustainably through music. Students might critically examine their local places (urban, suburban, and rural) to identify how the musics they learn outside of school are locally or globally oriented; identify what this music says about living sustainably where they live; and challenge the ethical value of musics and practices that are unsustainable. This is a reorientation of musics and society—each is a reflection of the other.

I recognize that a possible paradox exists when studying indigenous cultures *as* ecologically sustainable. While many (maybe even most) indigenous groups live sustainably, it is unlikely that all do. North American Paleo-Indians' arrival in the Americas, about 11,500 years ago, coincided with the extinction of large mammals, such as mammoths and mastodons (Burney and Flannery 2005; Fisher 1984; Grayson and Meltzer 2002). These groups seem to have influenced the distribution of trees (Foster et al. 2004), creation of grasslands, and erosion of soil (Denevan 1992). The global megafauna extinction is increasingly blamed on human activities. I can't help but see parallels between then and today. The unprecedented species loss we experience in our century is called the sixth great extinction (Wilson 2016). Facing the sixth great extinction, teachers and students would do well to look at all practices they intend to bring into the school, and identify those that are most life sustaining.

Many indigenous groups today live on the forefront of environmental activism. The indigenous environmental movement provides a unique and integrated approach to environmental justice in the United States (Clark 2002; Schlosberg and Carruthers 2010). With a shared experience of being invaded and relocated, the movement challenges ideas of land use, capitalism, and ecological sustainability, creating a counterhegemonic alliance against mining and manufacturing (Clark 2002). Further, in the fight for ecological sustainability, indigenous groups embrace diverse notions of environmental justice, modeling "a community-based, capabilities-centered conception of environmental justice" (Schlosberg and Carruthers 2010: 30).

Along with humility, problematizing institutions, and recognizing paradoxes, cultural diversity seems to typify the *philosophy on soil* metaphor, as soil is made up of diverse and interconnected species of bacteria, fungi, algae, worms, insects, and mammals. Soil is a local and diverse community.[3] All of these "living organisms are active agents" (Carson 2002: 54) in the soil community. The soil community is a powerful metaphor for music education that advances as decentralized, diverse, agentic, and interdependent.

The soil produces music. Though many of the musics of soil agents are inaudible to humans, some are not—the music of insects, such as termites tapping, and crickets scraping (Schafer 1994: 35). Perhaps the most thought-provoking inaudible sound humans experience is the sound of the big bang. Even today, the imprint in the way galaxies are clustered mirrors the intense sound waves created by the big bang. An audible model was first produced in 2003 (Cramer 2013). Even though it is not a sound we can hear every day, it is fascinating to reflect that our daily life is accompanied by a tone that began at the dawn of time. This inaudible drone underlines the audible musics of streams, waves, and wind—and of us.

Music educators' musics often displace local musical cultures, labeling young people as *non-musicians* (and a small group as *musicians*), and reifies a hierarchy of musical genres (see Bates 2011; Shevock 2016). Further, through our institutions, children learn to value distant places and musics, becoming uprooted from well-known soil (Prakash and Esteva 2008; Prakash and Stuchul 2010; Shevock 2016). But we are bound to the soil, bound to actual places and foods grown and musics performed by people we can see, hear, and touch. To re-root music education *on soil* will mean reassessing values, critically reconsidering music education history, and changing practices from: large to small; centralized to decentralized; and universal to local. Rooted musicking practices might replicate what Devall and Sessions (1985) called "Earth-bonding rituals, celebrating specific places" (27). Teachers can begin by recognizing "the kinds of ignorance produced by schools" (Prakash and Stuchul 2010: 68), listening to rooted musical cultures (beginning with urban, rural, and indigenous peoples—the social majorities) for sustainable ways for musicking well on soil.

Creative Vignette

He begins the general music class by telling a story.[4]

> My dad used to take me to the same fishing hole each spring on the first day of trout. We would talk as we walked down the railroad tracks in the near dark of early morning. Turning left at a point with no notable markers, we hiked through high grass, one in front of the other. When we approached the sandy bank, we got quiet. We climbed down a small incline to the water. We put worms to hook and checked our watches. The sun slowly arose through the gnarly and ancient trees and fog floated above the water. The creek flowed slowly here. A thick bush hung over the dark water. For what seemed like the longest time, we sat and waited. Trout season started at eight. If you wanted to crowd near the dam, you could show up fifteen minutes early. But then you'd spend the morning shoulder to shoulder, untangling your line. I learned from a young age, our "private" hole was valuable enough to leave before dawn. There was plenty of time to sit and listen to the early morning birdsongs (e.g., ducks, crows, sparrows). My favorite, black-capped chickadees, repeatedly sung their descending whole steps, "cheeseburger." When it was time, we cast our lines into the creek. Every time I hear the chickadees singing their descending song, I remember the first day of trout. I am there, sitting on a cold stump, baiting my line, and waiting for eight. I wouldn't trade a minute of that time for the world.

Handing out the lead sheet, he plays a composition for the class he wrote inspired by this experience.[5] Explaining his composition process, he found an online recording of black-capped chickadee calls on Wikipedia's Commons. Sitting at the piano, listening to the recording, he improvised a chord progression and, later, a melody. Then, using GarageBand, he added synthesizer sounds, producing an introduction ("on tape") that would lead in the composed and improvised sections.

Students begin reflecting on nature sounds, and compose their own music.

Music Educating in Place

Living on soil is living in place with human and non-human animals, geographies, climates, and cultures. Actual places are not easily generalized. As the old saying goes, the map is not the territory. Place-conscious pedagogy has recently been considered in music education.[6] Stauffer (2009) makes a distinction between *space*, which has to do with geography, and *place*, connected to lived experiences in a space. This distinction is useful in cultivating a philosophy on soil, because it is people's lived experiences (or a lack of those experiences), cultivating ecological consciousness, that are the starting point for eco-literate music pedagogy (Shevock 2015).

Music educators have noted that place-based and place-conscious pedagogies can benefit students in rural places (Bates 2011; Brook 2016; Prest 2016; Shevock 2016). It has been argued that, because "they are small enough to be managed democratically and humanely" (Theobald 1997: 161), rural places might provide the best opportunity for "healing" the education profession's errors. Music education, for rural students, can lead to feelings of being out of place. This out-of-placeness can be understood culturally/musically, and physically. But rural places can be understood as valuable resources for school musics (Bates 2011; Corbett 2016; Shevock 2016). As Corbett suggests, "First of all, a good music curriculum will help rural youth understand that their Uncle Ernie is a legitimate musical producer" (25).

Place-conscious pedagogy can be inclusive and relevant to students. Prest (2016) noted, "A place-conscious and relational approach to music education that engenders musical knowledge while also fostering bridging relationships within the classroom and community becomes more relevant and inclusive of more people" (152). Eco-literate music pedagogy is *for* people living in urban, suburban, and rural places. Place-conscious opportunities abound (exploring urban gardening, urban soundscapes, etc.), even in the most densely populated cities. And within their endless grass deserts, monocultures of the worst suburbs are in a constant battle with diversification of species and cultures. Through conscientization and work, all places can become more ecologically sustainable.

Music education scholars Benedict (2007) and Bates (2011) in the U.S., and Schmidt (2014), analyzing Brazilian NGOs, utilized Gruenewald's (2003) place-conscious critical pedagogy, in which "Place-based education, with its emphasis on local solutions and environmental awareness is combined with a critical pedagogy that encourages and enables students to interrogate contexts, along with multiple forms of oppression, in which they live" (Bates 2011: 118). A critical pedagogy of place opens space for teachers and students to interrogate the intersections of culture and ecosystems. A critical pedagogy of place is harmonious with philosophy on soil to the extent to which ecological justice is a critical issue that arises in the classroom, and where local places, the places of people living in "the margin" (Benedict 2007: 34), are respected and loved.

Little research is done on place-based music education. Some research, even while using place theory, uses a conceptualization of *place* inconsistent with a philosophy on soil.[7] Brook's (2016) case study research of a British Columbia, Canada music education program noted that place-based pedagogy can strengthen students' connection to in-school, local, and provincial communities. VanDeusen's (2016) case study explored the perceived value of a music program in the Midwest, U.S.A., by community and school members. Themes that arose were community history; school commitment to providing a comprehensive education to students; and music teacher interest in and openness to the community. Spring (2016) used narrative methodology to study four rural music educators in Ontario, Canada. They viewed their programs as enhancing the relationship between the school and communities, enhancing a sense of pride as well as recognizing responsibilities toward global communities. It is notable that this research is qualitative. It may be that place-based practices are best studied on small scales.

In a similar manner to the qualitative music education research, Gray and Thomson's (2016) mixed-methods study looked at an arts-based approach (including music) generally in their case study of Touched by the Earth, a yearlong Australian enrichment program. Students had conversations with "artists, scientists, historians, and Aboriginal elders," which served as an "immersion in the place as a creative stimulus for film, dance, poetry, writing, drawing, and music" (240). A music education relevant task shared in this write-up was a group project: students created a video of themselves "having fun" in nature, including music. The researchers found that "love and connection with the Earth are central to promoting deep engagement and indelible learning" (255).

The Commons

The commons can provide music educators with an opportunity to understand the ecological crises. Defined as "the environment that is available for use by the entire community," these commons encompass "forms of aesthetic expression" (Bowers 2006: 2) including musics (Bates 2016). Enclosure of the commons (privatizing and monetizing previously shared resources) has been considered one of the most relevant challenges today.

Developing ecological literacy begins with cultivating ecological consciousness in local places (Shevock 2015). People, though not all people, have been living and musicking sustainably in their commons for many generations. Consider the place-bound wisdom of indigenous peoples fighting ecological degradation today (mentioned earlier in this chapter). Global aims unresponsive to local understandings can be catastrophic for environments.

In contrast, indigenous groups have lived sustainably for many generations, rejecting short-term profit if it means long-term pollution. Additionally, enclosure, when considered with Schafer's (1994) soundscape, provides an interesting point of critical analysis. "The silence that is the background of natural sounds, is . . . being impacted (enclosed) by the spread of industrial culture" (Bowers 2006: 5). Daily, in cities, suburbs, and rural places, beautiful soundscapes are enclosed for private profit. This enclosure parallels ecological destruction. It is often the loudest, least beautiful machines doing the most ecological damage, and it is the loss of untouched wilderness that most contributes to loss of species diversity, and the resultant diversity of the musics of birds, amphibians, insects, and so on. Soundscape diversity may or may not be a problem to be addressed by music educators, but there seems to be an opening owing to understandings of timbre, texture, and other musical elements musicians consider.

Enclosure leads to a loss of both cultural and ecological diversity (Bowers 2006). Enclosure is characterized by privatization, displacement, standardization, stratification, and degradation (Bates 2016: 11). Analysis of music in the age of global industrial economy begins with an analysis of this loss of diversity. Bowers (2006) points out, "The diversity of the world's cultural knowledge systems that influence whether the local commons will continue to exist on a sustainable basis is now being threatened by the youth who are turning away from the intergenerational knowledge of their communities" (139). Musics the industrial economy produces represent an unsustainable monoculture (fewer genres, fewer musicians). *Musician* is redefined as professional musician (leaving the community musician as obsolete). And "we have rendered the average citizens of the country politically impotent" (Theobald 1997: 160). *Music* becomes something produced elsewhere (or maybe even nowhere), as children are taught to be *helpless* as musicians (Bates 2016), and *need* to be elsewhere—uprooted modern individuals (Shevock 2016). Even without the music education industry, the music industry educates children every day, transforming "people into interchangeable morsels of population, ruled by the laws of scarcity" (Illich 1990: 1).

Because the ecological crises are global, local understandings are the beginning, but not the end, of cultivating ecological literacy. Particular challenges to cultivating ecologically sustainable local places involve "transboundary" problems such as climate change, pollution, and deforestation (Dietz et al. 2003: 1907). Dietz et al. suggest small-scale institutions have an advantage in understanding commons governance. However, the challenge to human survival is "to develop and deploy understanding of large-scale commons governance quickly enough to avoid the large-scale tragedies that will otherwise ensue" (1910). They recommend adaptive governance, allocating authority "at multiple levels from local to global (nesting)" (1910).

This type of approach seems appropriate for those commons where local people have a history and claims to the land and water. Sustainable policy becomes especially challenging when we begin to consider crises in those commons where people do not live (e.g., tourism in Antarctica; whale over-fishing in much of the South Pacific). This approach might leave Illich's (1990) conception of a philosophy on soil "something vastly different from managing behavior on a shared planet" (1), but not necessarily. Ecological literacy must still foster "the earthy virtue, self-limitation" (1), leaving open local, grounded approaches to self-limitation. A philosophy on soil provides the grounded approach to thinking about ecological crises—a type of gardening. Respecting the wisdoms of other people in other places leads to ecological literacy and can improve our chance at survival through wise collective action.

The social majorities (the majority of the world's population who the industrial economy would choose to leave behind; the "undeveloped," "undereducated," "non-musicians") are currently revitalizing their commons (Bowers 2006; Esteva 2015; Prakash and Esteva 2008). This revitalization might be well embodied in Esteva's (2015) active verb, *commoning*—"reclaiming and regenerating commons and creating new commons beyond the dominant economic and political system" (87). Commoning provides a challenge for music educators. Where previous generations of music educators have been involved in enclosure (Bates 2016), music educators today might re-common musics, musicians, and the places where musics happen.

Because it is only in its infancy, a full expression of what music education on soil will look like is uncertain. However, the recent qualitative research in place-based music education seems to point toward a humble practice (VanDeusen 2016) focused on connections to local, regional, and global responsibilities (Brook 2016; Prest 2016; Spring 2016). One substantial detail arises. In cultures untouched by music education, it seems everyone musics (sings, dances, or plays instruments—sometimes with gender-specific roles). Commoning may begin with reconnecting people with their right to be musicians: "music is something everyone can do, you don't have to be helpless, you can do things even if you aren't a specialist" (Bates 2016: 16).

Music education on soil begins on actual soil. Like growing a garden in a local place. Unlike global, generalizable knowledge, local places are not places of theoretical perfection. "The garden is an unhappy place for the perfectionist. Too much stands beyond our control here, and the only thing we can absolutely count on is eventual catastrophe. . . . Among the many, many things the green thumb knows is the consolation of the compost pile"

(Pollan 1991: 132). In music education on soil, we ask deep questions, such as what type of education will sustain life. And we look to lived practices (cases of people living sustainably in urban, suburban, and rural places) for inspiration. Like gardeners, because we stand on soil, we must be willing to relegate unsustainable practices to the compost pile, where, through the magic of Mother Nature, even the worst weeds become food for next year's life. Because soil is living, diverse, and integrated, in the next chapter ecological literacy will be theorized as interdisciplinary.

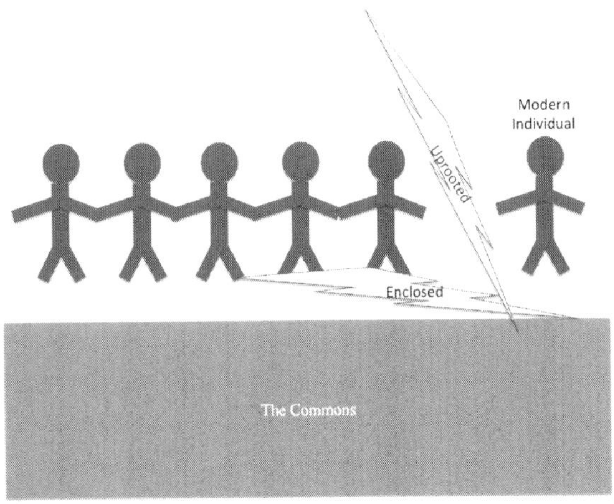

Figure 1.1 The Modern Individual, Enclosed and Uprooted

A Haiku

contemplating. me-
    -ditating. where do you close
and do I open?

Notes

1 From "Declaration on Soil" (Illich 1990: 1).
2 Hanh (1985) writes, "Walking meditation is practicing meditation while walking. It can bring you joy and peace while you practice it. Take short steps in complete relaxation; go slowly with a smile on your lips, with your heart open to an experience of peace. You can feel truly at ease with yourself" (1). "The Infant Buddha is often portrayed taking his first seven steps on earth, causing a lotus flower to appear in each of his footsteps. We should all cause a lotus flower to bloom with each of our peaceful steps" (5).
3 As Rachel Carson (2002) wrote, the soil is a "community [that] consists of a web of interwoven lives, each in some way related to the others—the living creatures depending on the soil, but the soil in turn a vital element of the earth only so long as this community within it flourishes" (56).
4 This vignette is reconstructed from childhood memories, and incorporates my jazz praxis (including a piece I composed as a musical artist in residence at Penn State Altoona), and how it is influencing my teaching pedagogy.
5 Link: https://youtu.be/UpWzFi3cqfY.
6 As an example, instrument construction can help students live well and connect to their environments (see Coleman 1922). In his ethnographic research, Matsunobu (2009, 2013) recommends place-based musical thinking. Describing a place-based practice, Japanese *shakuhachi*, the author suggests it benefits both spiritual and ecological wellbeing. Constructing flutes for performance is a central part of the practice: "musical thinking takes place when harvesting bamboo" (Matsunobu 2009: 94). Matsunobu (2013) makes a connection between instrument construction in *shakuhachi* and instrument construction in the pedagogy of Satis Coleman. "Her attempt at highlighting the intimate relationship between 'hands and mind' through instrument making may also be of great contribution to the field of music education" (191).
7 For instance Koops's (2014) case study utilizes place-based education theorists like Gruenewald and Stauffer to analyze the family *car seat* as a "place." While this provides an interesting data analysis, the speedy nature of vehicles is incongruent with the conceptualization of place utilized in music education on soil. This version of place-based education "sets itself apart from the mobility promised by mainstream education" (Prakash and Stuchul 2010: 67).

References

Abrokwaa, Clemente K. 1999. "Indigenous Music Education in Africa." In *What Is Indigenous Knowledge? Voices from the Academy*, edited by L. M. Semali and J. L. Kincheloe, 191–207. New York: Falmer Press.

Bates, Vincent C. 2011. "Sustainable School Music for Poor, White, Rural Students." *Action, Criticism, and Theory for Music Education* 12, no. 1 (April): 100–127.

——. 2016. "Toward a Sociology of Music Curriculum Integration." *Action, Criticism, and Theory for Music Education* 15, no. 3 (June): 8–20.

Benedict, Cathy. 2007. "Naming Our Reality: Negotiating and Creating Meaning in the Margin." *Philosophy of Music Education Review* 15, no. 1 (Spring): 23–35.

Bowers, C. A. 2006. *Revitalizing the Commons: Cultural and Educational Sites of Resistance and Affirmation*. Lanham, MD: Lexington Books.

Brook, Julia. 2016. "Place-Based Music Education: A Case Study of a Rural Canadian School." *Action, Criticism, and Theory for Music Education* 15, no. 4 (August): 104–126. DOI: 10.22176/act15.4.104.

Burney, David A., and Timothy F. Flannery. 2005. "Fifty Millennia of Catastrophic Extinctions after Human Contact." *Trends in Ecology and Evolution* 20, no. 7 (July): 395–401.

Carson, Rachel. 2002. *Silent Spring*. Boston, MA: Mariner Books. [Originally published in 1962.]

Clark, Brett. 2002. "The Indigenous Environmental Movement in the United States." *Organization and Environment* 15, no. 4 (December): 410–442.

Coleman, Satis N. 1922. *Creative Music for Children*. New York: G. P. Putnam's Sons.

Corbett, Michael. 2016. "Music Education and/in Rural Social Space: Making Space for Musical Diversity beyond the City." *Action, Criticism, and Theory for Music Education* 15, no. 4 (August): 12–29. DOI: 10.22176/act15.4.12.

Cramer, John G. 2013. "The Sound of the Big Bang." Online: http://faculty.washington.edu/jcramer/BBSound.html.

Denevan, William M. 1992. "The Pristine Myth: The Landscape of the Americas in 1492." *Annals of the Association of American Geographers* 82, no. 3: 368–385.

Devall, Bill, and George Sessions. 1985. *Deep Ecology: Living as if Nature Mattered*. Salt Lake City, UT: Gibbs M. Smith.

Dietz, Thomas, Elinor Ostrom, and Paul C. Stern. 2003. "The Struggle to Govern the Commons." *Science* 302, no. 12 (December): 1907–1912.

Esteva, Gustavo. 2015. "Time to Enclose the Enclosers with Marx and Illich." *International Journal of Illich Studies* 4, no. 1: 70–96.

Fisher, Daniel C. 1984. "Taphonomic Analysis of Late Pleistocene Mastodon Occurrences: Evidence of Butchery by North American Paleo-Indians." *Paleobiology* 10, no. 3 (Summer): 338–357.

Foster, H. Thomas, II, Bryan Black, and Marc D. Abrams. 2004. "A Witness Tree Analysis of the Effects of Native American Indians on the Pre-European Settlement Forests in East-Central Alabama." *Human Ecology* 32, no. 1 (February): 27–47.

Gray, Tonia, and Cameron Thomson. 2016. "Transforming Environmental Awareness of Students through the Arts and Place-Based Pedagogies." *LEARNing Landscapes* 9, no. 2 (Spring): 239–260.

Grayson, Donald K., and David J. Meltzer. 2002. "Clovis Hunting and Large Mammal Extinction: A Critical Review of the Evidence." *Journal of World Prehistory* 16, no. 4 (December): 313–359.

Gruenewald, David A. 2003. "The Best of Both Worlds: A Critical Pedagogy of Place." *Educational Researcher* 32, no. 4: 3–12.

Hanh, Thich Nhat. 1985. "From 'A Guide to Walking Meditation.'" Online: http://www.abuddhistlibrary.com.

Illich, Ivan. 1970. *Deschooling Society*. London: Marion Boyars.

——. 1990. "Declaration on Soil: A Joint Statement, Drafted in Hebenshausen, Germany, December 6, 1990, in Collaboration with Sigmar Groeneveld, Lee Hoinacki and Other Friends." Online: http://davidtinapple.com/illich/1990_declaraion_soil.PDF.

——. 2012. *Celebration of Awareness: A Call for Institutional Reform*. London: Marion Boyars. [Originally published in 1971.]

Koops, Lisa Huisman. 2014. "Songs from the Car Seat: Exploring the Early Childhood Music-Making Place of the Family Vehicle." *Journal of Research in Music Education* 62, no. 1: 52–65. DOI: 10.1177/0022429413520007.

Matsunobu, Koji. 2009. "Artful Encounters with Nature: Ecological and Spiritual Dimensions of Music Learning." PhD dissertation, University of Illinois at Urbana-Champaign. 3406782.

——. 2013. "Instrument-Making as Music-Making: An Ethnographic Study of *Shakuhachi* Students' Learning Experiences." *International Journal of Music Education* 31, no. 2: 190–201. DOI: 10.1177/0255761413486858.

Pollan, Michael. 1991. *Second Nature: A Gardener's Education.* New York: Grove Press.

Prakash, Madhu Suri, and Gustavo Esteva. 2008. *Escaping Education: Living as Learning within Grassroots Cultures*, 2nd edition. New York: Peter Lang.

Prakash, Madhu Suri, and Dana Stuchul. 2010. "McEducation Marginalized: Multiverse of Learning-Living in Grassroots Commons." *Educational Studies* 36, no. 1 (September): 58–73. DOI: 10.1207/s15326993es3601_6.

Prest, Anita. 2016. "Social Capital as a Framework for Music Education Research." *Action, Criticism, and Theory for Music Education* 15, no. 4 (August): 127–160. DOI: 10.22176/act15.4.127.

Schafer, R. Murray. 1994. *The Soundscape: Our Sonic Environment and the Tuning of the World.* Rochester, VT: Destiny Books. [Originally published in 1977.]

Schlosberg, David, and David Carruthers. 2010. "Indigenous Struggles, Environmental Justice, and Community Capabilities." *Global Environmental Politics* 10, no. 4 (November): 12–35.

Schmidt, Patrick. 2014. "NGOs as a Framework for an Education in and through Music: Is the Third Sector Viable?" *International Journal of Music Education* 32, no. 1: 31–52. DOI: 10.1177/0255761413488707.

Shevock, Daniel J. 2015. "The Possibility of Eco-Literate Music Pedagogy." *TOPICS for Music Education Praxis* 2015, no. 1 (October): 1–23.

——. 2016. "Music Educated and Uprooted: My Story of Rurality, Whiteness, Musicing, and Teaching." *Action, Criticism, and Theory for Music Education* 15, no. 4 (August): 30–55. DOI: 10.22176/act15.4.30.

Spring, Janet. 2016. "The Power of Metaphor in Rural Music Education Research." *Action, Criticism, and Theory for Music Education* 15, no. 4 (August): 76–103. DOI: 10.22176/act15.4.76.

Stauffer, Sandra L. 2009. "Placing Curriculum in Music." In *Music Education for Changing Times: Guiding Visions for Practice*, edited by T. A. Regelski and J. T. Gates, 175–186. New York: Springer. DOI: 10.1007/978-90-481-2700-0_14.

Theobald, Paul. 1997. *Teaching the Commons: Place, Pride, and the Renewal of Community.* Boulder, CO: Westview Press.

VanDeusen, Andrea. 2016. "'It Really Comes Down to the Community': A Case Study of a Rural School Music Program." *Action, Criticism, and Theory for Music Education* 15, no. 4 (August): 56–75. DOI: 10.22176/act15.4.56.

Wilson, Edward O. 2016. *Half-Earth: Our Planet's Fight for Life.* New York: Liverlight Publishing.

2 Ecological Literacy

> I was reminded, this morning before I rose, of those undescribed ambrosial mornings of summer which I can remember, when a thousand birds were heard gently twittering and ushering in the light, like the argument to a new canto of an epic and heroic poem. The serenity, the infinite promise, of such a morning! The song or twitter of birds drips from the leaves like dew. Then there was something divine and immortal in our life, when I have waked up on my couch in the woods and seen the day dawning, and heard the twittering of the birds.
>
> (Henry David Thoreau, 1852[1])

Teaching the Music of the Environment

Having completed a section on the elements of music (pitch, rhythm, melody, harmony, timbre, texture, dynamics, form, etc.), I challenged the Intro to Western Music students to write a listening response—an ethnographic/sonic description of place. Students were to find a place on campus with natural sounds (such as birds, squirrels, wind) and human sounds (such as footsteps, conversations, machines), and imagine the sounds they hear as purposeful; that they are listening to a music composition that warrants musical analysis. During the next class, with listening responses in hand, I prompted conversation with the following questions.

> What rhythms do you hear? How do the rhythms intersect and layer upon one another? Are the rhythms at various tempi?
>
> What types of pitches do you hear? Do any form short or longer melodic lines? Do various pitches harmonize?
>
> How is timbre expressed? Are there contrasts between harsh and smooth timbres?
>
> Is there a tempo to this composition? Is there a form? Dynamics? Texture?
>
> How does "the composer" coordinate the elements of music to yield a coherent composition?

Students replied using their listening responses as guides.

> "I was surprised by how much music was going on. I could hear a few different birds—tweet, tweet, tweet—outside my dorm room, through the window. That provides a high-pitched, unmelodic contrast to the low-pitched rise and fall of indistinct conversation just outside the door."
>
> "I also heard conversation and considered it a song . . . a freestyle rap. While pairs of people wait for each other to respond, multiple pairs discuss their days over each other creating a thick texture. I tried to transcribe some of the 'lyrics,' but it was impossible."
>
> "I sat by the pond. I pass that bench every day and hadn't noticed the melodies of the geese or the rhythms of the noisy squirrels. There was also an incessant drum roll sound—a large machine, of some sort, connected to the bookstore. Unlike the geese and squirrel songs, this machine song didn't rise or fall. It was . . . unnatural. Especially in comparison to the sounds of wild animals."

I prompted students with the question, "Considering this experience, what is 'natural' and 'unnatural'? And what does it mean to be a wild animal?" The discussion moved to defining these concepts—naturalness and wildness. Students disagreed with one another. Some considered all non-human and non-machine sounds wild and/or natural; and others teased out the difference between the concepts. These questions were guided by what Clark (2014) labels *deep ecological questioning*: "questioning about what 'wild' and 'domesticated' mean for ecosystems and places, communities and cultures, for minds and spirits, for you and me" (160). One benefit of deep ecological questioning is that it leads not only to respect for nature, but also to greater freedom.

> It leads us to a deeper conception of freedom that goes far beyond the conventional negative, "freedom from," or "doing what I want to do" kind of pseudo-freedom. Free minds and free spirits are not really like well-conditioned American middle-class adolescents, but more like wild nature—rich and diverse, spontaneous and creative, good and beautiful, at once orderly and chaotic. Free communities and free cultures have these same qualities of wildness. . . . To be really meaningful and transformative, the good and the sacred cannot be some distant ideal; we must be able to find them all around us, all the time, in our own lives and those of our families, friends and communities, in the creativity and self-expression of our culture, and at all levels of nature.
>
> (Clark 2014: 160–161)

Deep ecological questioning provides access to musicking as freeing and diverse; wild and respectful; and creative and in-place (not distant). I was using deep ecological questioning to re-root my teaching praxis to the environments where my students live. Finally, I provoked the students with the question: "What sounds did you contribute to nature's symphony?" None of the students had described their own sounds. Because the word "ecological" recognizes people as inseparable from the "environment," I wonder if my pedagogy is leading to an emerging "environmental" but not "ecological" consciousness. I challenge myself to do better.[2]

Creative Vignette

At an elementary school adjoining a small park, a veteran fifth grade science teacher leads her students on a walk. Treading familiar ground, teacher and students walk outside to listen to and identify the early spring birdsongs. Walking with notepad in hand, a student excitedly asks, "What type of bird is creating that noisy rhythm?"

Thinking musically, like an organically developing years-long symphony, the thickly textured, short chirps of house sparrows have become the park's dominant melody; and the upward glissandos of cardinals, the siren-like song of orioles, and the talking "heys" of grackles have somewhat faded to accompaniment. "That's one of the most common birds, the house sparrow."

She uses this experience as an opportunity to discuss the introduction of house sparrows to North America in the mid-19th century and this species' success in living alongside humans. Her students show interest in these birds that fly right up to them. When the class returns to the schoolroom, using their interest in house sparrows as a springboard, the teacher leads her students in a conversation about invasive species like house sparrows, European starlings, wild boars, and Burmese pythons.

A Passage

Nature regularly exhibits an abundance of musicking, from the songs of birds, to rustling leaves, to cicadas' recurring summer song (see Chapter 4, "Deep Ecology," for my defense of using the term *music* for human and non-human sound and meaning making). Educationally, children benefit when this sonic soundscape becomes a prominent part of their learning. From Rachel Carson's seminal *Silent Spring* to contemporary soundscape ecology (e.g. Pijanowski et al. 2011), sound has been a way scientists understand the health of the environment. Music teachers and students benefit by being connected to the places they live. As the Roman philosopher Seneca (4 BCE–65 CE) famously said, "all art is but imitation of nature." The music of nature is an immeasurable source of inspiration for musicking of all kinds. It also seems that nature's music provides inspiration in other arts (visual arts, drama, creative writing) and creative sciences, and can broaden opportunities for interdisciplinary-minded teachers of all subjects. In this chapter, I draw from interdisciplinary arts and ecological literacy literature to recommend a connecting approach to eco-literate music pedagogy.

A Provocation—Can/Should/Will We Intone Ecological Literacy?

What is literacy? Gould (2009) suggests there are four types of literacy: "functional, cultural, progressive, and critical" (47). But what is ecological literacy? If functional literacy (learning basic skills such as note reading) is nearly always present in music education, and critical literacy is "apparently missing" (48), what does that say for ecological literacy, which goes beyond transforming the social world to caring for the natural world? Ecological literacy can guide people to know when and when not to transform the natural world—it is Promethean or Epimethean when the challenge calls for either approach: transformation or conservation. Should we teach and learn music for ecological literacy? If so, why and how? Is it enough that ecological literacy gains the same footing as critical literacy: "increasingly becoming a high priority for some music educators, but . . . fairly meaningless to many others" (48–49)?

What do literacy and ecological literacy have to do with musicking? We seem to have a "shortcoming" when it comes to "the way we define knowledge" (Orr 1992: xi), including musical knowledge. Do centuries of nature inspiring music, and decades of musician activists supporting ecological movements count for nothing in the music classroom (see Chapter 3, "Ecomusicology")? "Ecological literacy . . . requires the more demanding capacity to observe nature with insight, a merger of landscape and mindscape" (86): and soundscape! What do people need to know for sustainable musicking? Perhaps we need to manage our musicking—"Management . . . more akin to childproofing a day-care center than it is piloting 'spaceship earth'" (162). Are we not culpable for ecologically destructive teaching/buying/consuming practices in music education?

Definitions of musicking might be, perhaps unintentionally, anthropocentric (e.g. Elliott and Silverman 2014; Small 1998). In this literature, musicking is defined as something *people* do. Is it *people* or *doing* that is important? For eco-literate music pedagogy, I suggest it is the *doing* (and sometimes not doing—not taking action) of music that takes precedence. What is important about *doing* in musicking? There are two aspects: 1) the intentionality of the musicker (e.g., the performer, composer, improviser, listener) selecting sounds for some purpose (e.g., communication, emotion, religious ritual); and 2) the experiencing done by the musicker(s), including listeners. Following Small (1998), it is in these relationships that the meanings lie; and in eco-literate music pedagogy, the intender(s) and/or experiencer(s) might be human or non-human. For instance, the ways wolves, whales, and songbirds music for communication, spiritual joy (there's an industry of meditation music using non-human musicking, often blended with synthesized human musicking), and art are also ways humans music for other humans. Additionally, non-human animals and humans music for each other (imitating each other's calls) in our connected ecosystem—and when doing so they use so-called "elements of music," creating what are experienced as works of art. Here I purposefully define *works of art* loosely, because it seems like egotism to declare another musicker's musicking to not be art without having a full understanding of their embodied intelligence. This humility is inspired by Nagel's (1974) famous conclusion of subjectivity in phenomenological understandings of non-human life:

> They are subjective, however, in the sense that even this objective ascription to experience is possible only for someone sufficiently similar to the object of ascription to be able to adopt his point of view—to understand the ascription in the first person as well as in the third, so to speak. (442)

We, frankly, do not share enough similarities with bats and other non-humans to make a statement about their understanding of artwork or spirituality; and it would be anthropocentric egotism (to use George Santayana's terminology) to claim these experiences only for humans. Further, it is dangerous to place humans above other fully evolved (all) non-human animals, because it reduces that life to abuses (see the work on *Animal Liberation* of Peter Singer, and Chapter 4, "Deep Ecology"). Therefore, I define music as *the intentional experiencing of sound*, and leave whether the intender and experiencer are human to the particular occurrence.

What are the responsibilities of music teachers and interdisciplinary teachers who include musicking activities when it comes to cultivating ecological literacy? What does ecological literacy through an interdisciplinary lens mean? What competencies are needed to enact music for ecological literacy in a classroom? What institutional barriers are built into 21st-century schools (public or private) when it comes to educating people for ecological literacy? These are the types of questions the eco-literate music pedagogue confronts on a daily basis. No single answer can be composed; and because of the nature of education—a creative project people enact together—a single, perpetual answer would be undesirable. Even for questions where seemingly easy answers can be found, the multiplicity of the ecological crises can complicate matters.[3] Rather, answers should be tentative, growing, co-constructed, reflected upon, enacted, and revised in actual classroom spaces.

When considering musicking in classrooms, the word *ecology* seems more insightful than *environment*. Students are part of their natural ecology. All people are. Environment seems to encourage a *self/other* relationship to nature that seems unjustified. "The term *environment* refers to the *external* conditions or surroundings of organisms, whereas *ecology* refers to the *relationships* between organisms and their external conditions or surroundings, that is, their environment" (Fox 1995: 8, emphasis in original). Here we can consider deep ecologist Warwick Fox's use of the term *relationships* with musicologist Christopher Small's (1998): "The act of musicking establishes in the place where it is happening a set of relationships, and it is in those relationships that the meaning of the act lies" (13). Musicking and music education may be best understood within the context of relationships—human-to-human, human-to-nature, and nature-to-nature relationships. Humans are natural animals, and human activity is natural activity. That understanding does not mean that human activity is always sustainable, but that musicking can and should be ecologically sustainable. In other words, music educators (and all educators), as part of an industrial enterprise (education), must consider the three postmodern educational "Rs" (reduce, reuse, and recycle) when making purchases. They must also consider whether their classroom affirms unsustainable consumer society or challenges it.

The word *literacy* has come to mean more than mere reading and writing (of words or music notation). For instance, in his 1960s adult literacy program in Recife, Brazil, Paulo Freire (2011) conceptualized literacy in connection to encouraging critical consciousness, which involves student creativity:

> From the beginning, we rejected the hypothesis of a purely mechanistic literacy program and considered the problem of teaching adults how to read in relation to the awakening of their consciousness. We wished to design a project in which we would attempt to move from naïveté to a critical attitude at the same time we taught reading. We wanted a literacy program which would be an introduction to the democratization of culture, a program with men as its Subjects rather than as patient recipients, a program which itself would be an act of creation, capable of releasing other creative acts, one in which students would develop the impatience and vivacity which characterize search and invention. (38–39)

For Freire and other critical pedagogues, teaching for literacy is about helping students developing critical consciousness of social, racial, and gender issues in their lives. Literacy, at least in the sense of reading and writing, is a means to an end—greater consciousness. But more than "consciousness of," that is acceptance of "mechanistic society, mechanical thinking, material reality" (Allsup 2003: 162). Literacy becomes a tool students use to create and change their worlds. Literacy also becomes a metaphor for that process of conscientization. To be "literate" is re-weighted to mean being critically conscious. Critical pedagogy requires dialogue (e.g., conversation), which empowers students as "Subjects" empowered to make sense (constructionism) of the objects (writing, music, etc.) in their lives. Freire's work has been highly influential, initially in adult education and, since, in other educational disciplines, including music. Music literacy becomes an act of creation for students. However, Freire never conceptualized his pedagogy for ecological literacy; rather, ecological literacy was conceptualized by other theorists.

Ecological literacy begins when teachers ask themselves: *What does a citizen need to know about the ecological crises*[4] *in order to make informed decisions?* (see McBride et al. 2013; Orr 1992). It is concerned with developing "the key ecological knowledge necessary for informed decision-making, acquired through scientific inquiry and systems thinking" (McBride et al. 2013: 3). Dedicated to creating sustainable human societies, eco-literacy has drawn heavily on the work of environmental studies professor and activist David Orr. For Orr (1992), relevant and sustainable education must begin with "*an uncompromising commitment to life and its preservation.* Anything less is morally indefensible" (133, emphasis in original). This provides a minimum bar for ethical music education: a commitment to preserve (human and non-human) life. Orr defines this commitment: "By commitment to life I mean a commitment, pervading learning and research at all levels, to health, harmony, balance, wholeness, and diversity as these qualities apply to both human and natural systems" (133). If literacy is loosely described as capacity to read, and musicianship as capacity to music, ecological literacy involves asking "What then?"

> The failure to develop ecological literacy is a sin of omission and of commission. . . . By failing to include ecological perspectives in any number of subjects, students are taught that ecology is unimportant for history, politics, economics, society, and so forth. And through television they learn that the earth is theirs for the taking. The result is a generation of ecological yahoos without a clue why the color of the water in their rivers is relevant to their food supply, or why storms are becoming more severe as the planet warms. (85–86)

Ecological literacy is imperative and interdisciplinary, and recognizing our shortcomings in long-term, integrated ways is key. "This failure [to recognize our shortcomings] is reflected in that portion of our sciences, humanities, and social sciences that deal with (or ignore) the relation between humanity and the earth. Ultimately, the failure can be traced to our schools and to our proudest universities" (xiii). With the paucity of articles dealing with music and the ecological crises in music education's research and practitioner journals, it seems fair to inculpate music education in this failing. To make amends for this failing in the face of the ecological crises, eco-literate music pedagogy can be a note toward atonement: Can/should/will we intone ecological literacy?

Interdisciplinary Education

As described in the Prelude, eco-literate music pedagogy involves cultivating ecological consciousness in local places, and connecting local understandings to global ecological crises (Shevock 2015). In other words, cultivating a love of place and recognition of the ecological crises are the means and ends of eco-literate music pedagogy. Cultivating ecological literacy in educational institutions seems to be an interdisciplinary task (Orr 1992; Shevock 2015); and many eco-literacy frameworks are employed in the fields of environmental education, ecology, and the humanities (McBride et al. 2013). Interdisciplinary education has been defined "as connecting distinct disciplinary concepts and facets of inquiry to create a common understanding of a complex issue, question, or problem" (Younker and Bracken 2015: 39). The terms *interdisciplinary* and *integration* are often conflated. Previously, I pointed to Liora Bresler's (1995) arts integration styles as a starting point for considering interdisciplinary education for ecological literacy (Shevock 2015: 7–8). Currently, I see Julie Thompson Klein and Janet Barrett's conceptions of *interdisciplinarity* as particularly appropriate.

In the U.S. and elsewhere, interdisciplinarity is frequently valued in general education (Klein 1990). Interdisciplinarity is "rooted in the ideas of unity and synthesis, evoking a common epistemology of convergence" (11). In eco-literate music pedagogy, interdisciplinarity emerges as a "social demand," that is, "Particular needs and new subjects which cannot, by definition, be contained within a single disciplinary frame, such as environmental research" (41). Eco-literate music pedagogy is interdisciplinary, in part, because the unnatural separation of disciplines has often caused music educators to focus on sonic aspects of music, forgetting the social and ecological inspirations connected to musicking (see Bates 2016). However, "*A truly comprehensive music program is already interdisciplinary in nature because musical understanding draws upon many forms of knowing and understanding*" (Barrett et al. 1997: 20, emphasis in original).

Interdisciplinarity emerges as a liberal process of connective thinking, and in music education can involve intradisciplinary, interdisciplinary, and student-centered/emancipatory practices (Barrett 2016: 173). In Barrett's model, *intradisciplinary* refers to connections made between various musical genres and styles. An example of intradisciplinary eco-literate music pedagogy might involve guiding students to understand musical eco-activism by comparing and contrasting Hunter Hensley's "Requiem for the Mountains,"[5] Mike Townsend's "They Call It Progress,"[6] and Will.i.am's "S.O.S."[7] *Interdisciplinary* refers to connections made between closely related disciplines. In this case, students might research ecological content (science), write poetry (writing), and compose melodies (music), composing their own eco-activism songs. Of course, songwriting is an interdisciplinary activity, where songwriters research, write, and music all at once: this follows the theory (found, for instance, in Deweyan education theory) that there is an underlying unity to interdisciplinary work (Klein 1990: 11). In the student-centered/emancipatory model, students make connections between music and their own experiences. In student-centered eco-literate music pedagogy, students could be assigned homework to ask their community elders to teach them songs about the environment, which they would then share in class.

Interdisciplinary education often locates its foundation in John Dewey's educational philosophy (Barrett 2016; Younker and Bracken 2015). Dewey (1975) taught, "There cannot be two sets of ethical principles, one for life in the school, and the other for life outside of the school" (7). Pragmatically speaking, we all live interdisciplinary, integrated, connected, complementary lives. "Interdisciplinarity and interactivity are not just buzz words; they are fundamental concepts that underpin how students think, how they learn, and how they will determine their futures" (Bassnett 2002: 108). All citizens of modern democracies must understand complex issues emerging from many disciplines, including the ecological crises. And yet in schools, too often, the disciplines are conceived as strictly separate; "although educational institutions segment knowledge into separate packages called 'subjects,' deep understanding often depends upon the intersections and interactions of the disciplines" (Barrett 2001: 27). The principle of complementarity suggests "the interplay of ideas [between/among disciplines] that fold over to strengthen bonds and deepen understandings" (Barrett 2016: 175). Eco-literate music pedagogy requires such interplay between music, environmental sciences, and other disciplines.

How might teachers enacting eco-literate music pedagogy choose repertoire? We can begin to answer this question by looking to Barrett's interdisciplinarity. Barrett (2016) recommends we consider music's interdisciplinary quotient (IQ)—"potential to lead fluidly to the study of closely related works, art forms, and eras" (177). High IQ music might be selected for various reasons in eco-literate music pedagogy. Students might sing David Mallett's "Garden Song" while growing food in class, learning how to care for soil, and reflecting on the ecological costs of shipping food around the world. Or the teacher might introduce Beethoven's "Pastoral" Symphony and challenge students to find parallels between Beethoven's work and gender metaphors such as *delicateness* and *purity* (or, in Beethoven's words, "the trees and the rocks give man the resonance he needs"); and following the insight of ecofeminists such as Karen Warren (2000), this conversation can lead to critical consciousness of the similarities among the unjustified historical domination by men of both women and nature.

Interdisciplinarity is used as a model for research in music education, but seldom includes ecological issues outright. Music education researchers Betty Anne Younker and Jillian Bracken (2015) shared an example of interdisciplinary education through birdsong (ecology and music). The purpose of their study was "to examine the challenges and effectiveness of a curriculum unit as implemented in a fourth-/fifth-grade general and music classroom" (38). Their pedagogy utilized Deweyan inquiry-based learning, and was developed concurrently by a music educator, an ecologist, and a science educator across four schools in two different U.S. states. Over a twelve-lesson unit focused on group composition, students learned to aurally identify the musical elements (timbre, pitch, and duration) of various birdsongs. Students were led through the following activities: realizing birdsongs on different instruments; studying how birdsongs function in the environment; function as communication; exploring melodic ideas, texture, and timbre; and deciding on a musical form for their compositions. Their research concluded that the birdsong group work allowed for improved student social engagement while solving problems.

To sum up, using an interdisciplinary approach a teacher recognizes knowledge, issues, and problems as connective (between disciplines); because of this, an interdisciplinary approach seems essential to eco-literate music pedagogy. Interdisciplinarity is fundamental to how people think, and music (connected to other disciplines) provides insight into ecological knowledge, issues, and problems. Music with a high IQ (subjectively estimated) can be selected to shed light on various ecological issues. While the concept "IQ" as drawn from Barrett (2016) and recycled from intelligence testing to consider repertoire might at its worst become mechanistic, if IQ for ecological literacy remains subjective, teachers should be able to avoid some of the failings of the racist and classist use of intelligence quotient. This interdisciplinary IQ is not a hierarchical arrangement of students into a preset theory—the bell curve. Finally, interdisciplinary music learning can help improve students' social engagement. With advantages of interdisciplinarity revealed, next I draw attention to one critical aim of music education: ecological literacy.

David Orr's Ecological Literacy

Ecological literacy is an interdisciplinary praxis. The possibility of musicking to cultivate ecological literacy rests on two principles: 1) it is possible to learn to understand the ecological crises and creatively imagine and realize solutions; and 2) music contributes uniquely to human ecological thinking. The concepts of musicking, values, and culture intersect (see Small 1998). Part of the problem of the ecological crises is cultural—and maybe all of it, which depends on the extent to which *culture* is distinguishable from other closely related concepts such as economy, history, technology, and religion. Discussing modern rootlessness, Orr (1992) suggests, "The anomie, rootlessness, and alienation of the modern world are part of a larger system of values, technologies, culture, and institutions which also produce acid rain, climate change, toxic wastes, terrorism, and nuclear bombs" (4). This is consistent with Weil's[8] (1952) treatise *The Need for Roots*, and Illich's (1990) *Philosophy on Soil*; and it has been noted that music education reinforces this "larger system of values" (e.g., Bates 2013a, 2013b, 2016; Shevock 2016). As Bates (2016) argued, "music curriculum integration ought to be down-to-earth—rooted in authentic (naturally occurring as opposed to technically constructed), complex, everyday musical practices and the immediate and long-term needs of children" (9). Cultivating ecological literacy involves the formation of attitudes, and involves directly experiencing nature, experiences that are rare in modern society generally and schooling places specifically (Orr 1992: 86–89).

One important aspect of David Orr's (1992) ecological literacy is his analysis of the ecological crises as plural. He identified three crises—the food crisis, the fuel crisis, and climate change. Each crisis involves many elements.

> The first is a food crisis evident in two curves that intersect in the not too distant future: one showing worldwide soil losses of twenty-four billion tons, the other a rapidly rising world population. The second crisis on the horizon is that caused by the end of the era of cheap energy. We are in a race between the exhaustion of fossil fuels, global warming, and the transition to a new era based on efficiency and solar energy. The third crisis, perhaps best symbolized by the looming prospect of a global climate change, has to do with ecological thresholds and the limits of natural systems. We can no longer assume that nature will be either bountiful or stable or that the earth will remain hospitable to civilization as we know it. These three crises feed upon each other. (3)

Intersecting with these three crises, more than twenty years since Orr's analysis, we might also consider crises connected to loss of species diversity (particularly the collapse of marine life due to overfishing), the continued loss of wilderness to development (e.g., rainforests in Brazil and Louisiana), lack of apex predators in wilderness preserves, environmental toxins (PCBs, pesticides, phthalates, VOCs, dioxins, heavy metals, chlorine) negatively affecting human and non-human life, oil spills, overpopulation of non-native species, nuclear meltdowns, loss of potable water, and desertification of arid regions.

Whether there are three or thirty ecological crises, the consequence of Orr's analysis that the crises we face are plural is vital to cultivating ecological literacy. We live in an age where political discourse hardly makes room for one crisis—climate change. Climate change is certainly an important global crisis, and considered by itself a glut of ways forward (none of which we seem to be attempting) can be introduced. But the realization that the ecological crises are plural demands we consider other crises when solving this one. For instance, some solutions involve increased production of plastics, which can exacerbate the waste crisis and increase our already untenable levels of e-waste production. This might add to the Great Pacific Garbage Patch and exacerbate the collapse of marine life. An ecologically literate citizenry will be crucial if effective solutions—both small- and large-scale—are to be imagined, enacted, and enforced in communities around the globe. These citizens "understand global issues [and] know how to live well in their places" (Orr 1992: 1). I have previously made the distinction between ecological consciousness (involving living well in a local place) and ecological literacy (being able to understand the global crises) (Shevock 2015). Thus musicking experiences cultivating local ecological consciousness and helping to conceptualize global ecological literacy involve listening and respecting local places as well as asking deep questions about global crises.

Deep questioning about the ecological crises is a central praxis of deep ecology (discussed more in Chapter 4, "Deep Ecology"). However, a little on respecting local places is appropriate here. Most often, modern people (who our educational systems educate) do not listen to local places. We rather prefer a universal, genderless, placeless consumer society that produces musics as universalizables, especially those appropriated musics of distant places. Music teachers often enter schools with predetermined repertoire, and this act shows little respect for the human and non-human musicking in local places, because it avoids "tradition, bounded by place . . . transform[ing] people into interchangeable morsels of population, ruled by the laws of scarcity" (Illich 1990: 1). Musickers (e.g., performers, composers) become scarce in industrial society, so that consuming becomes a manufactured need (in Illichan terminology, nonconvivial): manufactured and marketed by music educators. Lifelong musicking, then, becomes a process of teaching students to become lifelong consumers,[9] where in generations before music educators arrived (and in the social majorities around the world untouched by music education) everybody musicked, not for profit or fame, but in local places, sustainably.

Ecological Literacy in Practice

What does it mean to individual music teachers, each in unique circumstances, to teach music for ecological literacy? For example, what does it mean for me to teach music for ecological literacy as a university lecturer? How do I balance the general spirit of this introductory course in Western music with reflection on us in the class as part of Western society; as expressions of that society; as flawed, willful, individualistic, and career-oriented; as unsustainable; in short, occupied in what Western philosophy calls the "human condition." How can I respect the students' selves (as they currently understand themselves) while nudging them (and myself) to imagine their view of self might include unknown responsibilities for perpetuating the mindsets leading to the ecological crises? It is a difficult task, to stare into the void and realize that some changes are easy and others are hard; but it is essential to see change as possible—that ecological literacy is, in practice, initially about self-change.

Self-change begins with self-understanding, with respecting our personal histories and using the places we came from as stable starting points, walking into a sustainable future. My own early life experiences, growing up in a small, rural, Central Pennsylvania town, may have influenced my trajectory toward eco-literate music pedagogy. My grandfather was active in building both the town park and the nearby Prince Gallitzin State Park. As a child, I spent long hours in the woods, hunting or fishing with my father and uncle, and later by myself. My father always kept two gardens, a backyard garden for food and a frontyard flower garden around our statue of Mary. I helped him turn soil each year. These early experiences with nature led to my lifelong love of nature. It was only natural that, as an adult music educator, I would find eco-literate music pedagogy alluring—hoping to cultivate in my students a similar (but unique to their own experience) appreciation for the musicking of nature.

Interdisciplinary—Self-Reflective—Place-Based—Constructivist

Ecological literacy practices are vital to multiple disciplines, include teachers scrutinizing the sustainability of their classrooms, are rooted in place, and involve constructing knowledge. A large body of scholarship uses David Orr's ecological literacy in understanding learning (e.g., Gruenewald 2003; Jordan et al. 2009; Prakash 1995; Rigolon 2012; Taylor 1997). Taylor (1997) used art education to understand the problem of waste and sustainability. Brewer (2002) considered the use of schoolyards as laboratories to cultivate ecological literacy. Waters (2011) experimented with experimental play in natural environments (using found objects as artifacts) to guide students to compose "anthropomorphic narratives," and to challenge their anthropocentric attitudes. Uncovering solutions to the ecological crises will involve interdisciplinarity, respect for traditional ways of thinking about how people and nature relate, self-reflection, and self-cultivation. Many researchers have noted the importance of early experiences to cultivating ecological consciousness (Hulmes 2013; Kilcup 2009; Rigolon 2012), and the importance of place to that process (Gruenewald 2003; Rigolon 2012; Waters 2011) and ecological action (Jordan et al. 2009).

Douglas Forell Hulmes (2013) explored his personal history, and how his early memories of Scandinavian heritage led to his becoming an environmental educator. An early appreciation for Nordic mythology, which includes how the people relate to nature, led to his appreciation of how traditional stories lead to wisdom and awareness of how to live ethically within the natural world. His autoethnography concluded that landscape influences culture and that, while "students consider relationships between subjects that are often taught in isolation" (para. 6), "each holds a key that together unlock the door to our future . . . one key will not open the door to a sustainable future for human kind" (para. 32).

Ecological literacy is interdisciplinary, self-reflective and critical, place-based, and constructivist. Hulmes's (2013) autoethnography shows how early experiences and rooted mythology can lead to student interest. Living in an unsustainable consumerist society, students must be interested in the condition of nature if we hope they'll appreciate ecology and become passionate, sustainable citizens. This focus on interest and appreciation leads me back to experience—all people benefit from musical engagement with nature.

Musicking for Ecological Literacy[10]

On a warm April morning, I walk through Tudek Park in Pennsylvania (U.S.A.) with my eight-month-old son. We begin our walk, and the primary sounds we hear are those of children singing their joyous, rhythmic songs; songs of excited sliding and swinging on playground equipment, and sounds of playing soccer; children running and kicking the black and white ball into a small net. The pieces of play equipment have their own sounds—squeaking and swishing. There are also the sounds of adults talking; a family has a reunion under a pavilion's wall-less rooftop. Adult conversation and childish shouts intermingle. The rumble of cars and the singing of birds underlie these sounds, but are present nonetheless. My aim lies deeper in the park, since my child is too young to use the playground. As we walk down the hill, the sounds of birds and squirrels become more prominent, while human sounds fade. Dogs are barking further into the park, and we sit, shoeless, in the grass under a newly budding tulip tree, somewhat (but not completely) away from human sounds.

Here we listen to the music of the park, and I try to sing some of the sounds I hear, and my son also sings in a low monotone. We touch the dirt, grass, the tree's trunk, branches, and the buds emerging from this tree as we crawl on the ground and sing. It's possible that many people walking by would not call our songs music, though everybody (mostly dog walkers) stops to say "Hi" to my son, and he sings back his low droning song ("da, ah, ah, da"), often crescendoing at the end for emphasis. The experience of nature's music is a learning experience—for me and my son—where my aim is to cultivate ecological consciousness: feeling a sense of place, of the environment, of non-human life (the ecology in which we live).

Listening

In humility, people music together in complex, interdisciplinary ways. All acts of musicking involve more than mere sounds. In place, musicking can occur in sustainable ways, to learn together about our ecology, and to live together with human and non-human musickers. It seems easy and natural to enact eco-literate music pedagogy outside of schooling institutions. But can we, with our consumption-driven music education traditions, have experiences that parallel those some parents have with their children? Can we listen and music sustainably? Can we cultivate ecological literacy? It seems there is a path teachers might take.

In this chapter I considered the meaning of ecological literacy to eco-literate music pedagogy. An interdisciplinary approach to cultivating ecological literacy, rooted in place, was recommended. The next chapter will listen more deeply to what ecomusicology can add to eco-literate music pedagogy.

listen-waiting[11]

connected, linking, joining, thinking
integrating,
    listen-waiting,
nature thrums, hums, pulsating
beaks as drums, cicada trumpets
jumping, playing,
    listen-waiting,
disciplines descend through thought
children bringing what they brought
squirrels, their heavy, raucous dancing
resonant and playful prancing
    listen-waiting, listen-waiting
writing, reading, science speeding
slowing, dawdling, singing, modeling
novel ways of understanding,
nature's melodies in phrases,
connected, linking, joining, thinking
integrating,
    listen-waiting

Notes

1 This opening quotation is taken from the highly influential Sierra Club publication: Eliot Porter, ed., *In Wilderness Is the Preservation of the World* (New York: Ballantine Books, 1967), 42.

2 In Chapter 4, "Deep Ecology," I delve more into the deep ecology movement, and how the platform and ethics of that movement have guided my pedagogy.

3 Take as one example the choice to move to a paperless music classroom. At first glance, it seems an easy choice—go paperless. If the teacher is doing so to lower her classroom's carbon footprint, it becomes a matter of simple mathematics—how much CO_2e (carbon dioxide equivalent) does each option generate in the long run? The average carbon footprint of a paperback book is 1 kg CO_2e (Berners-Lee 2011: 77); and the average e-reader such as an iPad has approximately 50 kg CO_2e (78)—if you are planning on using more than fifty books during the life of the iPad, the iPad is the better choice. However, the question becomes more complicated when you consider a second ecological crisis, such as waste (waste and carbon footprint seem like the most reasonable aims in going paperless). Hoping to concurrently reduce e-waste, you consider that e-waste is a distinctive global crisis. The total amount of yearly e-waste is approximately 41.8 Mt (megatonne; or 41.8 million metric tons), and about 15.5 percent of e-waste, 6.5 Mt, was reported as formally treated (recycled or safely disposed of) (Bladé et al. 2015). Capitalism, our dominant social ideology, makes realizing long-term sustainability even more difficult to measure. With the rapid commercial release of new generations of iPads (first generation in March 2010; Wi-Fi and Wi-Fi + 3G versions released in April 2010; iPad 2 in March 2011; iPad 3 in March 2012; iPad 4 in October 2012; iPad Mini in November 2012; iPad Air in October 2013; iPad Mini 2 in November 2013; iPad Air 2 and iPad Mini 3 in October 2014; and iPad Pro in September 2015), the ecological footprint of an up-to-date paperless classroom becomes unsustainable. Alternatively, a music teacher must take into account amount of use; the other teachers who will use these iPads; how iPads will be purchased (if they're already in the school); and if there are unique creative pedagogies (e.g., iPad ensembles using popular music pedagogy) that will be employed providing more inclusive opportunities for more students—hence the importance of educating the next generation of educators for ecological literacy. It is entirely possible that solutions to one ecological crisis might exacerbate another. The plurality of crises can lead to arts teachers feeling "overwhelmed" (Taylor 1997: 16). An eco-literate music teacher is prepared to make decisions for his/her classroom.

4 Even Thoreau was responding to a crisis—the crisis of civilization.

> I wish to speak a word for Nature, for absolute freedom and wildness, as contrasted with a freedom and culture merely civil—to regard man as an inhabitant, or a part and parcel of Nature, rather than a member of society. I wish to make an extreme statement, if so I may make an emphatic one, for there are enough champions of civilization: the minister and the school committee and every one of you will take care of that.
>
> (Thoreau 1851: n.p.).

These words for the opening of his essay "Walking" emphasize the crisis of civilization Thoreau was motivated by. More than a century later, Americans would find themselves hard pressed to find commons to occupy (as in *Walden*; Thoreau 1995) or be able to merely walk westward into wilderness without being viewed as trespassers on someone's property. The crisis of civilization is, after all, interlinked with the population crisis; and there are many more humans living in my America than in Thoreau's.

5 Link: https://youtu.be/g8gJrTe2O18.
6 Link: https://youtu.be/50v-nckcoSg.
7 Link: https://youtu.be/_05lX4V7YY4.
8

> A human being has roots by virtue of his real, active and natural participation in the life of a community which preserves in a living shape certain particular treasures of the past and certain particular expectations for the future. This participation is a natural one, in the sense that it is automatically brought about by place, conditions of birth, profession and social surroundings.
> (Weil 1952: 43).

In Weil's definition of *roots*, where she describes the condition of post-World War II Europe, we see the concerns articulated that eventually emerge as ecological literacy, place-based education, and eco-literate music pedagogy.

9 In some music education scholarship, "independent music literacy" is understood threefold as composing/improvising, performing, and "*responding to music* with understanding, as an informed consumer or audience member" (Reimer 2011: 27). Eco-literate music pedagogy challenges the ends of all three of these aspects of literacy, and re-focuses us on sustainability.
10 This section is drawn from my publication "The Possibility of Eco-Literate Music Pedagogy" (Shevock 2015).
11 Poetry is one way to artfully elicit, explain, and illustrate understandings in the autoethnographic tradition (see Chang 2008: 148).

References

Allsup, Randall Everett. 2003. "Praxis and the Possible: Thoughts on the Writings of Maxine Greene and Paulo Freire." *Philosophy of Music Education Review* 11, no. 2 (Fall): 157–168.

Barrett, Janet R. 2001. "Interdisciplinary Work and Musical Integrity." *Music Educators Journal* 87, no. 5 (March): 27–31. DOI: 10.2307/3399705.

——. 2016. "Adopting an Interdisciplinary Approach to General Music." In *Teaching General Music: Approaches, Issues, and Viewpoints*, edited by Carlos R. Abril and Brent M. Gault, 168–182. New York: Oxford University Press.

Barrett, Janet R., Claire W. McCoy, and Kari K. Veblen. 1997. *Sound Ways of Knowing: Music in the Interdisciplinary Curriculum*. New York: Schirmer Books.

Bassnett, Susan. 2002. "Is There Hope for the Humanities in the 21st Century?" *Arts and Humanities in Higher Education* 1, no. 1 (June): 101–110. DOI: 10.1177/1474022202001001008.

Bates, Vincent C. 2013a. "Drawing from Rural Ideals for Sustainable School Music." *Action, Criticism, and Theory for Music Education* 12, no. 1 (April): 24–46.

——. 2013b. "Music Education Unplugged." *Action, Criticism, and Theory for Music Education* 12, no. 2 (September): 75–90.

——. 2016. "Toward a Sociology of Music Curriculum Integration." *Action, Criticism, and Theory for Music Education* 15, no. 3 (June): 8–20.

Berners-Lee, Mike. 2011. *How Bad Are Bananas? The Carbon Footprint of Everything*. Vancouver: Greystone Books.

Bladé, C. P., F. Wang, R. Kuehr, and J. Huisman. 2015. *The Global E-Waste Monitor—2014*. United Nations University, IAS—SCYCLE. Bonn, Germany. Online: http://i.unu.edu/media/unu.edu/news/52624/UNU-1stGlobal-E-Waste-Monitor-2014-large-optimized.pdf.

Bresler, Liora. 1995. "The Subservient, Co-Equal, Affective, and Social Integration Styles and Their Implications for the Arts." *Arts Education Policy Review* 95, no. 5 (May–June): 31–37.

Brewer, Carol. 2002. "Conservation Education Partnerships in Schoolyard Laboratories: A Call Back to Action." *Conservation Biology* 16, no. 3 (June): 577–579.

Chang, Heewon. 2008. *Autoethnography as Method*. Walnut Creek, CA: Left Coast Press.

Clark, John P. 2014. "What Is Living in Deep Ecology?" *Trumpeter* 30, no. 2: 157–183.

Dewey, John. 1975. *Moral Principles in Education*. Carbondale: Southern Illinois University Press. [Originally published in 1909.]

Elliott, David J., and Marissa Silverman. 2014. *Music Matters: A Philosophy of Music Education*, 2nd edition. New York: Oxford University Press.

Fox, Warwick. 1995. *Toward a Transpersonal Ecology: Developing New Foundations for Environmentalism*. Albany: State University of New York Press.

Freire, Paulo. 2011. *Education for Critical Consciousness*. New York: Continuum. [Originally published in 1974.]

Gould, Elizabeth. 2009. "Music Education Desire(ing): Language, Literacy, and Lieder." *Philosophy of Music Education Review* 17, no. 1 (Spring): 41–55.

Gruenewald, David A. 2003. "The Best of Both Worlds: A Critical Pedagogy of Place." *Educational Researcher* 32, no. 4: 3–12. DOI: 10.3102/0013189X032004003.

Hulmes, Douglas Forell. 2013. "An Autoethnographic Exploration of Norway: Nature and Culture." *Journal of Sustainability Education* 5 (May). Online: http://www.jsedimensions.org/wordpress/content/2013/05/.

Illich, Ivan. 1990. "Declaration on Soil: A Joint Statement, Drafted in Hebenshausen, Germany, December 6, 1990, in Collaboration with Sigmar Groeneveld, Lee Hoinacki and Other Friends." Online: http://davidtinapple.com/illich/1990_declaraion_soil.PDF.

Jordan, Rebecca, Frederick Singer, John Vaughan, and Alan Berkowitz. 2009. "What Should Every Citizen Know about Ecology?" *Frontiers in Ecology and the Environment* 7, no. 9 (November): 495–500.

Kilcup, Karen L. 2009. "Fresh Leaves: Practicing Environmental Criticism." *PMLA* 124, no. 3 (May): 847–855.

Klein, Julie Thompson. 1990. *Interdisciplinarity: History, Theory, and Practice*. Detroit, MI: Wayne State University Press.

McBride, B. B., C. A. Brewer, A. R. Berkowitz, and W. T. Borrie. 2013. "Environmental Literacy, Ecological Literacy, Ecoliteracy: What Do We Mean and How Did We Get Here?" *Ecosphere* 4, no. 5 (May): 1–20. DOI: 10.1890/ES13-00075.1.

Nagel, Thomas. 1974. "What Is It Like to Be a Bat?" *Philosophical Review* 83, no. 4 (October): 435–450.

Orr, David W. 1992. *Ecological Literacy: Education and the Transition to a Postmodern World.* Albany: State University of New York Press.

Pijanowski, Bryan C., Luis J. Villanueva-Rivera, Sarah L. Dumyahn, Almo Farina, Bernie L. Krause, Brian M. Napoletano, Stuart H. Gage, and Nadia Pieretti. 2011. "Soundscape Ecology: The Science of Sound in the Landscape." *BioScience* 61, no. 3 (March): 203–216.

Porter, Eliot, ed. 1967. *In Wilderness Is the Preservation of the World.* New York: Ballantine Books. [Originally published in 1962.]

Prakash, Madhu Suri. 1995. "Ecological Literacy for Moral Virtue: Orr on [Moral] Education for Postmodern Sustainability." *Moral Education* 24, no. 1: 3–18.

Reimer, Bennett. 2011. "Professional Notes: President Shuler's Guiding Principles." *Music Educators Journal* 97, no. 4 (June): 27–28.

Rigolon, Alessandro. 2012. "A Greener Future: The Active Role of Place in Enhancing Ecoliteracy in Children." *Journal of Architectural Planning Research* 29, no. 3 (Autumn): 181–203.

Shevock, Daniel J. 2015. "The Possibility of Eco-Literate Music Pedagogy." *TOPICS for Music Education Praxis* 2015, no. 1 (October): 1–23.

——. 2016. "Music Educated and Uprooted: My Story of Rurality, Whiteness, Musicing, and Teaching." *Action, Criticism, and Theory for Music Education* 15, no. 4 (August): 30–55. DOI: 10.22176/act15.4.30.

Small, Christopher. 1998. *Musicking: The Meanings of Performing and Listening.* Middletown, CT: Wesleyan University Press.

Taylor, Pamela G. 1997. "It All Started with the Trash: Taking Steps toward Sustainable Art Education." *Art Education* 50, no. 2 (March): 13–18.

Thoreau, Henry David. 1851. "Walking." [Kindle DX version.]

——. 1995. *Walden, and On the Duty of Civil Disobedience.* Project Gutenberg. Online: http://www.gutenberg.org/ebooks/205. [Originally published in 1854.]

Warren, Karen J. 2000. *Ecofeminist Philosophy: A Western Perspective on What It Is and Why It Matters.* Lanham, MD: Rowman & Littlefield Publishers.

Waters, Philip. 2011. "Trees Talk: Are You Listening? Nature, Narrative and Children's Anthropocentric Place-Based Play." *Children, Youth and Environments* 21, no. 1: 243–252.

Weil, Simone, 1952. *The Need for Roots.* New York: Routledge.

Younker, Betty Anne, and Jillian Bracken. 2015. "Inquiry-Based Learning through Birdsong: An Interdisciplinary Project-Based Experience." *Journal of Music Teacher Education* 24, no. 3 (April): 37–52. DOI: 10.1177/1057083714527110.

3 Ecomusicology

> We need to see the Earth in its sequence of transformations as so many movements in a musical composition. The sequence of events that emerge in time needs to be understood simultaneously, as in music: the earlier notes are gone when the later notes are played, but the musical phrase, indeed the entire symphony, needs to be heard simultaneously. . . . Each new theme alters the meaning of the earlier themes and the entire composition. The opening theme resonates throughout all the later parts of the piece.
>
> (Thomas Berry, 1999[1])

Teaching Music as Ecological Activism

After studying the nature-inspired compositional techniques of Emily Doolittle, I guided the Intro to Western Music students through YouTube videos that typify some of the ways people music for ecological justice: Hunter Hensley's "Requiem for the Mountains,"[2] "They Call It Progress" by Mike Townsend,[3] Vancouver Opera violinist Carolyne Cole performing for protesters in British Columbia opposing Kinder Morgan's pipeline survey,[4] Will.i.am's "S.O.S.,"[5] and "Don't Frack My Mother" by Sean Lennon, Yoko Ono, and Artists Against Fracking.[6] In small groups, students discussed what each artist hoped to accomplish; whether each artist's aims were local or global; how classical and popular music activism relate to and differ from one another; and what these say about music in society and nature.

> "Both Hensley and Townsend hoped to stop mountaintop removal in Kentucky. Their aims were local. Similarly, Cole's playing was for a specific group of protesters, and Lennon and Ono's song aims to stop fracking in New York."
>
> "Will.i.am takes more of a global perspective."
>
> "That makes sense. He is more widely known than the other people we listened to."

"Many of the other artists seem to write about local issues, but don't deliberately connect them to global problems."

"S.O.S. asks for God's help. I wonder if that's because we cannot fix climate change without divine intervention."

"But he does point to greed, and that's something we can change. We can be less greedy."

"'Don't Frack My Mother' aims to raise awareness. Both of the Kentucky songs are about raising awareness. Cole aims to bring beauty or relief to protesters after a long, stressful day protesting the pipeline."

"The popular music styles had an advantage. Lyrics can be tailored to a specific cause. Hensley and Cole are using pre-existing music, and neither were written for today's problems."

"Does Hensley's use of a requiem, a mass for the dead, say anything?"

"Not much."

"Nobody speaks Latin anymore."

"It seemed a bit hopeless because these mountains are already gone, forever."

Eco-literate music pedagogy emerged as a confluence of personal experiences and reading (of ecological literacy, indigenous knowledge, and deep ecology). During this time, I became aware of ecomusicology, a new discipline that emerged during my doctoral studies and continues to grow. Ecomusicology is a subfield of both musicology and ethnomusicology and brings those disciplines' traditional focus together with ecocriticism. Though other humanistic disciplines have a longer history of studying nature, as recently as 2002 nature was described as "a fashionable postmodern topic" (Rehding 2002: 305) for musicologists. Composers like Beethoven (Rehding 2002), Messiaen (Hold 1971), R. Murray Schafer (1994), and Pauline Oliveros (2005) have long been influenced by nature's musics. In 2011 the *Journal of the American Musicological Society* published a special edition focused on ecomusicology. Ecomusicologists study nature's portrayal by composers, the musics of non-Western cultures, and music for ecological activism. Three recent books have helped define this emerging subfield (Allen and Dawe 2016; Pedelty 2012, 2016), providing insights into the intersections of environment and music, which helped inform the creation of eco-literate music pedagogy. Pedelty's two books focus on activism and sustainability in popular music, and Allen and Dawe edited a book that theorizes and shares examples of ecomusicology in non-Western cultures. In this chapter, I will describe ecomusicology in Western music, ethnomusicology, and popular music, and explore some possibilities for eco-literate music pedagogy.

Ecomusicology and Western Music

Ecomusicology locates its seminal writings in composer R. Murray Schafer's *soundscape*, work he began in the 1960s at Simon Fraser University.Schafer's World Soundscape Project endeavored to increase "public awareness of sound, documenting environmental sound and its changing character,[7] and establishing the concept and practice of soundscape design as an alternative to noise pollution" (Westerkamp et al. 2006). Because humans impact ecosystems (the current epoch—geologic period longer than an age—is often called the Anthropocene), soundscapes can be understood as "composed." Not only have environmental musics shaped compositions (e.g., Beethoven's "Pastoral" Symphony; Olivier Messiaen's *Réveil des oiseaux*; Emily Doolittle's *Falling Still*; Laurie Spiegel's *Anon a Mouse*[8]), but also human-made intentional (and sometimes frustratingly unintentional) sounds (e.g., cars, construction equipment, boom boxes, chatting) dominate the soundscape most of us experience daily. The discipline of ecomusicology emerges as a critical study of the intersections of music and environment (Allen 2011; Taylor and Hurley 2015), and current studies use Schafer's insights (Cook 2013; McCutchan 2011; Seeger 2016). Paradoxically, to study intersections assumes music and environment are separate, an assumption challenged deeply by music education on soil (see Chapter 4, "Deep Ecology"). Nonetheless, the current chapter functions as a literature review. Beginning from *intersections* seems to have allowed ecomusicology to emerge.

The musicological study of Messiaen's music may have helped move scholarship toward ecomusicology, because "One cannot discuss the music of Olivier Messiaen without mentioning bird-song" (Hold 1971: 113). To understand birdsong's influences on composers, researchers must consider "biological" and "aesthetic" (114) questions. Messiaen's transcriptions involve "imaginative moments" (122), transmuting birdsong transcriptions for artistic purposes. Messiaen seems to place value on the musics of birds, and *Réveil des oiseaux* has three dedications. These include dedications to an ornithologist and to "birds of our woods" (116). This suggests a scientific/ornithological and an artistic/ecocentric aim. This dual aim can be found in soundscape compositions and in contemporary ecomusicology literature. In eco-literate music pedagogy, music educators use this precedence to inspire an interdisciplinary approach, incorporating scientific and artistic aims.

Non-Human Agency

Messiaen's dedication appears to be aligned with Cook's (2013) analysis of agency in the first movement of George Crumb's *Ancient Voices of Children.*

"The song engaged with the world through a negotiation of identity, agency, territory, and survival; that is to say, the song enacts an acoustic, and specifically vocal, ecology" (126). Crumb's music accomplishes this by enacting "a familiar nocturnal soundscape and its vocal ecology: when crickets have voice, humans are mute, and vice versa" (127). Since music provides the listener (and other musickers) with an opportunity to construct meaning, especially when it comes to human relationships (see Small 1998), giving agency to crickets (even when not actually present) produces a way to think, through music, about our relationship to nature. It may be that non-human animals make music purposefully, teach and learn music, and can change music for their own reasons. Historically, Satis Coleman (1922) suggested some animals teach music to their young (101–102). More recent scholarship suggests, while some animals' musicking is instinctive, some is learned: including "among birds, parrots, hummingbirds, and oscine songbirds, and among mammals, humans, cetaceans, pinnipeds, elephants, and bats" (Doolittle and Gingras 2015: R820).[9]

Place

McCutchan's (2011) memoir shares Louisiana musician and naturalist Earl Robicheaux's soundscape music, which documents the environmental beauty and degradation of Atchafalaya, Louisiana, in North America's largest rainforest. This book is particularly interesting for eco-literate music pedagogy, as it is placed and localized (see Chapter 1, "Philosophy on Soil"): it draws on Robicheaux's extensive memories and recordings of Atchafalaya to describe and challenge environmental degradation. For instance, Robicheaux's soundscape compositions, in which he coordinates the musics of non-human animals, especially birds, are used as background for interviews of Atchafalaya residents and photographs and presented at museums. This work aims at "preserving Louisiana's culture and natural sound environments."[10]

The shift in musicology from studying environmental influences on the music, to music and the ecological crises parallels a similar shift in other humanistic disciplines. "Engagements with social concerns have now broadened to include a desire on the part of some musicologists (and others in the humanities) to respond to issues of climate change, sustainability, and environmental degradation" (Taylor and Hurley 2015: 1). McCutchan's memoir seems to fulfill Allen's (2011) description of ecomusicology as "emphasizing the importance of place" and evoking "critical and self-critical elements" (394).

For music educators, ecomusicology scholarship into Western music suggests a pedagogy that recognizes the enduring and varied inspiration nature has been for composers, performers, and listeners. It begins by recognizing that relationships are complex, and involve meanings intended by composers, performers, listeners, and others. Students can come to understand how understandings of human/nature relationships inspire certain musics (e.g., as pastoral or soundscape), and different musics encourage different relationships. It is important to consider that not all of these relationships are sustainable relationships—some are of domination (see Chapter 4, "Deep Ecology," especially the discussion around parallels of injustice). Students can also study current composers' intentions to change the way we think about those relationships.

Ecomusicology and Ethnomusicology

Many ecomusicologists study non-Western musical cultures. Studying non-Western cultures can provide insight into various ways people can understand human/nature relationships, and help Western music educators avoid ethnocentrism. The ways we conceive music and nature are culturally specific. Western cultures (especially the urbanormative,[11] an ideology dominating at universities) seem to reinforce a dichotomy between humankind and nature. But many cultures, including the Kĩsêdjê/Suyá of Brazil (Seeger 2016), the Yoreme of Mexico (Simonett 2016), Aboriginal Australians with music using the gumleaf (Ryan 2013), and Chinese with *yuanshengtai min'ge* (original ecology folksong) (Rees 2016), offer more of a relational perspective.

The Yoreme perceive "relational identities, solidarities, and obligations between human and non-human entities" (Simonett 2016: 105). Their nature is lived in, understood as it is experienced. Aboriginal gumleaf instruments exemplify the close relationship between musician and plant (Ryan 2013). Though "invented by educated urbanites" (Rees 2016: 54), *yuanshengtai min'ge* emphasizes the relationship of culture and environment. This genre, similar to many Western folksong traditions today, is environmentally conscious and aims to preserve diverse musical cultures. Experienced nature is essential to cultivating ecological consciousness, rooted in place and living on soil (Shevock 2015). Singing and using naturally occurring instruments (a practice originating in the 1920s with Satis Coleman's pedagogy) might be a good place to begin living on soil. Eco-literate music pedagogues can help in re-rooting music education to place, sustaining environments and cultures.

Nature

Seeger (2016) studied the Kĩsêdjê to demonstrate different concepts of nature and the musics of non-human animals. The Kĩsêdjê make an indefinite distinction between nature and village, and the ability to understand (e.g., which birdsongs signaled there wouldn't be good fishing) and imitate non-human animal sounds is highly prized. Seeger contrasts the Western privileging of eyes with the Kīsêdjê valuing of ears—"Instead of 'I see' they would say 'I hear'" (91). A Kīsêdjê song (most of their music is vocal) names particular species of fish. And to the Kīsêdjê, "Fish sing about themselves. Birds sing about themselves. Rodents (mice) sing about themselves. Bees sing about themselves and trees sing about themselves. The Kīsêdjê do not sing about themselves but instead sing the songs of other species" (94). This idea, that humans do not have music of their own, but rather are unique in their ability to sing of other species, provides a distinctive view of nature—people who are able to "obtain the animals' perspective" (95) are valued. For the eco-literate music pedagogue, echoing nearby non-human musics might help improve the ways we relate to nature, especially where Western culture–nature dichotomies dominate.

Because different cultures have different assumptions about human/nature relationships, ethnomusicology has the potential to provide music classrooms with radically different understandings of relationships through singing/performing/listening to music. Certain musics might help students gain non-human animals' perspectives, and develop solidarity with nature, crucial to becoming eco-literate citizens prepared to protect non-human life and conserve the environment. But conservation may not be enough (though ultimately I recommend a conserving based on deep ecology in the Postlude). Radical conscientization may be required (see Chapter 4, "Deep Ecology").

Creative Vignette

> Aurochs mooed more deeply than oceans \ in a dark, green forest, so histories say. \ And great auks swam so swiftly, in motion \ until our greed led their species away. \ Will Darwin's Frogs, or Golden Frogs, or Sacred Grove Frogs stop singing today? \ Brother life: Where do you start and I end? \ Sister life: Where do you start and I end? \ There must be time, at least let's pretend. \ Mother life: Where do you start and I end?

This year, the secondary English teacher felt gratified. Listening to the "musics" of endangered wildlife in class stimulated inspired student writing. Students read the final draft of their song lyrics aloud.[12] Some students wrote songs about family pets, or wildlife they regularly encountered. Many students were inspired by two poems read in class: Ogden Nash's "A Caution to Everybody" and Ciaran O'Driscoll's "Great Auks." These provoked lively classroom conversation about species collapse. The great auk, once a numerous species that features in both poems, became extinct in the 19th century, mostly owing to egg and specimen collection by humans. Today, amphibians face the greatest threat of extinction, and students discussed the critically endangered Chilean Darwin's frog and the Panamanian golden frog, and the extinct golden toad.[13] Many student lyrics were read about the trill of the eastern gray treefrog, a species that's abundant locally, to honor the amphibian soundscape in which they find themselves. Once-plentiful species are facing extinction because of human actions (e.g., unsustainable farming, suburbanization, spreading invasive species), and some scientists are calling our current Anthropocene epoch *the Sixth Extinction.* The teacher feels songwriting provides students with a way to know, to express feelings, and to enact a little bit of change in the face of horror.

Ecomusicology and Popular Music

Possibly the first environmental activist song, "Woodman! Spare that Tree!," written in 1837 by George Morris and Henry Russell, was connected to the American Transcendentalist movement (Kahn 2012). In 1966, Pete Seeger wrote the first album dedicated to environmental activism, *God Bless the Grass* (Ingram 2008). A number of popular music examples were discussed in Intro to Western Music, shared in the reflection (e.g., Will.i.am's "S.O.S."). Many popular musicians advocate environmental activism. For instance,

Figure 3.1 Penny Spies a Dodo at Carnegie Museum, Pittsburgh

Michael Jackson's music, including "Earth Song," addressed racism, poverty, and environmental degradation (Martin 2012). Hip-hop can be used to explore issues of "the city's industrial decline, environmental waste, and how citizens are responding to these issues" (Farrugia and Hay 2014: 2). Interdisciplinary artists like Bob Wiseman are using song along with other arts (puppet theatre, film, and comedy) to illuminate ecological concerns in a way that engages audiences (Mark 2014). In eco-literate music pedagogy, teachers can provide students with the opportunity to write songs that address environmental concerns. While musical experiences can help teaching through songwriting, by focusing on lyrics, a degree in music education is far from necessary to facilitate meaningful experiences. Western culture is one of the few (maybe only) in which *only* "musicians" perform music, and there's no reason to believe that our professionalization is sustainable (it was Illich's classic *Deschooling Society* that first challenged professional schooling).

Sustainable Practice

Popular music activities seem fundamentally unsustainable, and ecomusicologists studying popular musicking have contributed to the issue of sustainability. Unsustainable energy use presents a problem for large, global and local, participatory musicians. As Pedelty (2012) states, "rock and roll is energy-intensive" (3), and he calls the belief that local musical cultures are sustainable today "a romantic notion."

> Developing more sustainable ways of life is difficult in a world where culture is increasingly independent of place. Yet ecological knowledge and consciousness are essential, especially for those with economic power and privilege. Much of the world is already living modestly, but not by choice. It is important for us to become more aware of distant others who share our air, water, and ecosystem. (201)

Even if less energy-intensive popular musicking can be cultivated, economic interests challenge ecologically oriented musicians and audiences. For instance, after the BP oil spill, BP dedicated money to support Gulf Coast tourism, including music festivals (Pedelty 2012; Stimeling 2014). Stimeling studied several music projects connected to this effort. "BP's music tourism efforts might be seen as an effort to draw attention away from the spill's ongoing, long-term consequences" (11), and "potential visitors can be reassured that the spill was little more than a minor inconvenience that does not merit continued political action or judicial action" (13). It seems essential that teachers are aware of the intersectional challenges arising among ecology and economics. Achieving a sustainable praxis in our unsustainable society may be impossible. Perhaps the best we can achieve is slightly more sustainable. Perhaps that will be enough.

Musicking Ecological Activism

Ecological activism takes place locally. Even global activities like 350.org's global climate initiatives are organized locally (McKibben 2014). Activism is best a place-based practice. This local focus is paralleled in ecomusicology. Exemplifying this, Pedelty (2016) studied musicians and ecological activism along the Salish Sea, and Stimeling (2012) studied Central Appalachia.

Musicking activists draw upon nostalgia to stimulate change. The music written around mountaintop removal in Central Appalachia uses many genres, which bring different ideologies, and *nostalgia* is often, but not solely, connected to country, folk, and bluegrass genres. For instance, Captain Catfeesh's "Blair Mountain Battle March" draws on "leftist punk ideology" and a do-it-yourself ethos, trying to produce music "unmediated by capitalism" (Stimeling 2012: 18). Whether country, bluegrass, or punk, musicians evoke "a sense of nostalgia that is rooted in competing understandings of local history, identity, and landscape" (19). Adult educators may have the greatest opportunity to use this nostalgic response, but even our youngest students may have experienced environmental degradation. Mountains are being removed, rivers polluted, and trees cut for development every year.

Ecomusicologists studying popular music provide music educators with insight into ecological activism. Music can provide the opportunity for small acts of resistance. It may even be that, as Berry (1999) suggested, music cultivates a sophisticated way of thinking about living well on earth, understanding the environment as a "sequence of transformations" (27) with early notes still resonating. But this cannot be understood only as "music," but as "musics." A fundamental lesson drawn from ecomusicology is that different musics encourage different relationships between humans and humans, and humans and nature. The soil is made up of diverse beings, and so music classrooms should incorporate as much musical diversity as possible. This inclusion benefits music teachers, students, performers, and listeners as well as those agents perhaps not involved directly, such as non-human animals.

Sitting Beneath This Sycamore Tree

Sitting beneath this sycamore tree, engrossed in jay songs and hope
  neglecting the damp and the scent of decay
Indulgent children of cars and of bridges,
  but waking on bedrock and dirt
Surrounded by cell walls with no noticed scope
  a selfish debtor to nature's sure call
And wanting to make the wide world better
  but having one small tool to use
  and that one may be too obtuse
    music may be too abstruse
      neglecting the jay's firm "jay"
  and even so, let's let it loose

Notes

1 From ecotheologian Thomas Berry's (1999: 27) book *The Great Work*. This quotation points to a musical way of thinking about the planet, environment, and our ecological crises, conscious of history while living in the moment, something *good* (both ethically "good" and "effective") musicians do very well.
2 Link: https://youtu.be/g8gJrTe2O18.
3 Link: https://youtu.be/50v-nckcoSg.
4 Link: https://youtu.be/nX0iycbJ4QE.
5 Link: https://youtu.be/EfOlw8PvYVQ.
6 Link: https://youtu.be/8zeldweayF8.
7 This focus on the negatively changing quality of soundscapes aligns well with ecomusicology's focus on ecocriticism.
8 *Anon a Mouse*, interestingly, "explores deep ecological cross-species communication" (Feisst 2014: 21), placing humans and non-human animals on relatively equal musical footing.
9 Canadian composer Emily Doolittle's *zoomusicology* seems to parallel ecomusicology, but focuses on the musics of animals.
10 Link: http://louisianasoundscapes.com/.
11 Bates (2016) suggests that few scholars explore rural music education, which may be due to urbanormativity—privileging urban *centers* and treating rural places as the *periphery*—in universities.
12 Howard's (2012) research in Newfoundland found creative texts and music can be used to expand students' awareness of place in literacy curricula. Preservation of music and stories was also an important component in identity formation. "In this space, we make and remake ourselves. Our identity, beliefs, values and relationships are expressed and communicated to others through words, written and spoken, through art and music, through ritual and ceremony" (336).
13 Stuart et al. (2004) point out that models usually use habitat loss, human activity, and climate change to predict amphibian extinction rates, but that these models underestimate extinction rates. Captive breeding is currently the only available defense scientist have. "Unless these declines are quickly understood and reversed, hundreds of species of amphibian can be expected to go extinct over the next few decades" (3).

References

Allen, Aaron S. 2011. "Ecomusicology: Ecocriticism and Musicology." *Journal of the American Musicological Society* 64, no. 2: 391–394.

Allen, Aaron S., and Kevin Dawe, eds. 2016. *Current Directions in Ecomusicology: Music, Culture, Nature*. New York: Routledge.

Bates, Vincent C. 2016. "'Big City, Turn Me Loose and Set Me Free': A Critique of Music Education as Urbanormative." *Action, Criticism, and Theory for Music Education* 15, no. 4 (August): 161–177. DOI: 10.22176/act15.4.161.

Berry, Thomas. 1999. *The Great Work: Our Way into the Future*. New York: Bell Tower.

Coleman, Satis N. 1922. *Creative Music for Children*. New York: G. P. Putnam's Sons.

Cook, Robert C. 2013. "The Vocal Ecology of Crumb's Crickets." *Journal of the Society for American Music* 7, no. 2: 123–145. DOI: 10.1017/S1752196313000047.

Doolittle, Emily, and Bruno Gingras. 2015. "Quick Guide: Zoomusicology." *Current Biology* 25 (October): R819–R820.

Farrugia, Rebekah, and Kellie D. Hay. 2014. "The Politics and Place of a 'Legendary' Hip Hop Track in Detroit." *Music and Politics* 8, no. 2 (Summer): 1–26. DOI: 10.3998/mp.9460447.0008.203.

Feisst, Sabine. 2014. "Animal Ecologies: Laurie Spiegel's Musical Explorations of Urban Wildlife." *Social Alternative* 33, no. 1: 16–22.

Hold, Trevor. 1971. "Messiaen's Birds." *Music and Letters* 52, no. 2 (April): 113–122.

Howard, Patrick. 2012. "Educating the Imagination: Teaching for Sustainability through a Bioregional Literacy Curriculum." *Journal of Education for Sustainable Development* 6, no. 2: 327–340. DOI: 10.1177/0973408213475375.

Ingram, David. 2008. "'My Dirty Stream': Pete Seeger, American Folk Music, and Environmental Protest." *Popular Music and Society* 31, no. 1: 21–36. DOI: 10.1080/03007760601061456.

Kahn, Richard. 2012. "Environmental Activism in Music." In *Music in American Life: The Songs, Stories, Styles, and Stars That Shaped Our Culture*, edited by Jacqueline Edmondson. Online: www.academia.edu/1395159.

Mark, Andrew. 2014. "Refining *Uranium*: Bob Wiseman's Ecomusicological Puppetry." *Environmental Humanities* 4: 69–93.

Martin, Sylvia J. 2012. "The Roots and Routes of Michael Jackson's Global Identity." *Culture and Society* 49: 284–290. DOI: 10.1007/s12115-012-9550-z.

McCutchan, Ann. 2011. *River Music: An Atchafalaya Story*. College Station: Texas A&M University Press.

McKibben, Bill. 2014. *Oil and Honey: The Education of an Unlikely Activist*. New York: Times Books.

Oliveros, Pauline. 2005. *Deep Listening: A Composer's Sound Practice*. Lincoln, NE: iUniverse.

Pedelty, Mark. 2012. *Ecomusicology: Rock, Folk, and the Environment*. Philadelphia, PA: Temple University Press.

——. 2016. *A Song to Save the Salish Sea: Musical Performance as Environmental Activism*. Bloomington: Indiana University Press.

Rees, Helen. 2016. "Environmental Crisis, Culture Loss, and a New Musical Aesthetic: China's 'Original Ecology Folksongs' in Theory and Practice." *Ethnomusicology* 60, no. 1 (Winter): 53–88.

Rehding, Alexander. 2002. "Review: Eco-Musicology." *Journal of the Royal Musical Association* 127, no. 2: 305–320.

Ryan, Robin. 2013. "'Not Really a Musical Instrument': Locating the Gumleaf as Acoustic Actant and Environmental Icon." *Societies* no. 3: 224–242. DOI: 10.3390/soc3020224.

Schafer, R. Murray. 1994. *The Soundscape: Our Sonic Environment and the Tuning of the World*. Rochester, VT: Destiny Books. [Originally published in 1977.]

Seeger, Anthony. 2016. "Natural Species, Sounds, and Humans in Lowland South America: The Kĩsêdjê/Suyá, Their World, and the Nature of Their Musical Experience." In *Current Directions in Ecomusicology: Music, Culture, Nature*, edited by Aaron S. Allen and Kevin Dawe, 89–98. New York: Routledge.

Shevock, Daniel J. 2015. "The Possibility of Eco-Literate Music Pedagogy." *TOPICS for Music Education Praxis* 2015, no. 1 (October): 1–23.

Simonett, Helena. 2016. "Of Human and Non-Human Birds: Indigenous Music Making and Sentient Ecology in Northwestern Mexico." In *Current Directions in Ecomusicology: Music, Culture, Nature*, edited by Aaron S. Allen and Kevin Dawe, 99–108. New York: Routledge.

Small, Christopher. 1998. *Musicking: The Meanings of Performing and Listening*. Middletown, CT: Wesleyan University Press.

Stimeling, Travis D. 2012. "Music, Place, and Identity in the Central Appalachian Mountain Removal Mining Debate." *American Music* 30, no. 1 (Spring): 1–29.

——. 2014. "Music, Place, and Gulf Coast Tourism since the BP Oil Spill." *Music and Politics* 8, no. 2 (Summer): 1–20. DOI: 10.3998/mp.9460447.0008.202.

Stuart, Simon N., Janice S. Chanson, Neil A. Cox, Bruce E. Young, Ana S. L. Rodrigues, Debra L. Fischman, and Robert W. Waller. 2004. "Status and Trends of Amphibian Declines and Extinctions Worldwide." *Sciencexpress* 14 (October): 1–5. DOI: 10.1126/science.1103538.

Taylor, Hollis, and Andrew Hurley. 2015. "Music and Environment: Registering Contemporary Convergences." *Journal of Music Research* 6: 1–18.

Westerkamp, Hildegard, Adam P. Woog, Helmut Kallmann, and Barry Truax. 2006. "World Soundscape Project." Historica Canada. Online: http://www.thecanadianencyclopedia.com/en/article/world-soundscape-project/.

4 Deep Ecology

> Instead of a culture of abundance, profit-driven globalization creates cultures of exclusion, dispossession, and scarcity. In fact, globalization's transformation of all beings and resources into commodities robs diverse species and people of their rightful share of ecological, cultural, economic, and political space.
>
> (Vandana Shiva, 2005[1])

But Why?

"But what is 'wilderness'?"
"Anything people don't touch."

A Challenge

This twelve-year-old girl at the summer music camp summarized what musicologist Rehding (2002) called the "negative approach to nature" (306) that dominates much current thinking about ecology. Wilderness is defined by what it does not contain—humans. Such an approach to nature and wilderness provides a challenge for music educators, since "any nature associated with music is steeped in culture" (306), whether Schafer's soundscape, Emily Doolittle's *Falling Still*, or Pete Seeger's *God Bless the Grass*. Music and nature are easily conceived of as a subset of culture—humans music to express something about the natural world. And yet, subsuming *nature* to *culture* seems to miss something important. Might wilderness refer to non-human musics, the musicking of birds, insects, wolves, and whales having intrinsic value, that is, value not placed upon it by people? While some of nature and music can be categorized as a subset of culture (as humans' cultural understandings of nature), I suspect this student was, in a basic and commonsense way, correct. Wild musicking (which "people don't touch"), if such a concept can be imagined by our oh-so-human minds (how could we be otherwise, and yet humility challenges philosophy!), has an existence separate from culture—and culture often can degrade wild musicking, paralleling the way human actions often degrade wild ecosystems.

Broadly speaking, there are two approaches to conceiving *value* in environmental philosophy: anthropocentric and ecocentric (Fox 2000).[2] Respectively, these positions can be described as those attributing *use value* and those attributing *intrinsic value* to non-human life. With the ecological crises, it seems humans have grossly misused non-human life. This type of distinction between use value and intrinsic value does not seem to be present in music education scholarship.[3] In this chapter I recommend the intrinsic value of human and non-human being and musicking.

With a robust understanding of ecological literacy and ecomusicology, I deepen the current concept under analysis, music education on soil. The deep ecology movement and the anthropocentric detour are historical as well as theoretical. The deep ecology platform provides a basis for an expanded praxis of teaching and learning music. I tell the stories "The Good Samaritan" and "The Birth of Pecos Bill" to frame an ecocentric valuing of non-human musicking. The current argument suggests non-human beings' musics have intrinsic value, rather than use value.

Historically in music education, Satis Coleman recommended children learn to love nature by listening to nature's music. Thus Coleman's pedagogy can be a reconnecting endeavor: reconnecting to nature, communities, and ourselves (Shevock 2015). As I enacted eco-literate music pedagogy, I began to wonder about our field's underlying values. Coleman's music education philosophy valued connection, while many of today's value individualism. I have been inspired by many music educators' writings about the social value of music and music education, but as I cultivated eco-literate music pedagogy I began feeling something might be missing: specifically, something of the way we usually value in music education. If *social* is understood as *human*, might social value not be enough? On the other hand, with what we know about the social lives of ants, of wolves, of orcas, and of other non-human beings, can we broaden our conception of *social*? What benefits might an ecocentric position offer music educators?

The Emergence of the Deep Ecology Movement

The student's answer to the wilderness question, "Anything people don't touch," followed a series of questions inspired by deep ecology. Deep ecology is sometimes defined as "deep" (as opposed to "shallow") because of its deep questioning—its search for root causes underlying the ecological crises (Naess 2005). Bowers (2006) connects anthropocentrism to the dominant mythopoetic narrative: "The root metaphor of anthropocentrism leads to viewing environments as a resource—which means it is moral to exploit it to its fullest" (33). Our industrial, technological, informational society exploits the environment because we understand humans as disconnected or different from the environment. We even understand the environment and ourselves as resources (Shiva 2005). Scholars connected with the deep ecology movement have been on the forefront of environmental scholarship and activism, especially beginning in the 1980s (Fox 1995).[4] Deep questioning of ecological crises seems to lead people to action for change.

At its core, deep ecology is based on the shallow/deep distinction, which was first proposed in the 1970s by Norwegian philosopher Arne Naess. One aspect of the shallow/deep distinction is the issue of intrinsic value; shallow ecology attributes use value to non-human life, while deep ecology attributes intrinsic value to non-human life. Naess's (2005) conception of deep ecology begins with the acceptance of "inherent value" for all life; he writes, "The flourishing of human and non-human life on Earth has inherent value. The value of non-human life-forms is independent of the usefulness of the non-human world for human purposes" (68). What do humans gain when conceiving of non-human life having intrinsic value?

Berry (1995) positions humans as uniquely reflective; "the universe is able to reflect on and celebrate itself in a unique mode of conscious self-awareness" (8). In praxial education, reflection is intimately tied to action (Freire 2000), and "musical understanding exists to empower people to create and listen to musical products and events of one kind or another" (Elliott and Silverman 2015: 208).

This reflection might parallel Freire's (2000) conception of conscientization. Freire referred to conscientization (*conscientização*) as "learning to perceive social, political, and economic contradictions, and to take action against the oppressive elements of reality" (35). In 1970 when he wrote *Pedagogy of the Oppressed*, Freire was conscious of issues connected to social, political, and economic contradictions, but he was most certainly not ecologically conscious. Freire's thinking is anthropocentric, and he talks about animals as unable to construct their world (98–101). Not only is Freire's position false when we consider the reality of musics created by whales, coyotes, and birds, but it is ecologically unsustainable when we consider the history of human ecological destruction (often through misuse, remaking the world into unlivable spaces for humans and non-human animals). It is likely Freire brought his anthropocentric positions unreflectively from the humanistic philosophy he extends. It is also likely that this anthropocentric perspective kept him from realizing the depth of the ecological crises, and it has been left to his students to design an ecopedagogy. Nonetheless, the language of conscientization (as perceiving contradictions and taking action), a learning process guiding teachers–students and students–teachers from naïveté to awareness, solidarity, and action, is worthwhile when discussing value in eco-literate music pedagogy.

Conscientization connects perception to action. Perceiving non-human musicking as having intrinsic value might open music education to ecological action.

The Deep Ecology Platform

Deep ecologists proposed an eight-point platform to facilitate deeper conversation around ecological crises in environmental philosophy and action. This platform has been published multiple times, with slight variations, in Devall and Sessions's (1985) seminal text, online by the Foundation for Deep Ecology,[5] and by Arne Naess, the philosopher who coined the term *deep ecology*. The version shared here is in Naess (2005), and is modified from the original 1984 version (68).

1. The flourishing of human and non-human life on Earth has inherent value. The value of non-human life-forms is independent of the usefulness of the non-human world for human purposes.
2. Richness and diversity of life forms are also values in themselves and contribute to the flourishing of human and non-human life on Earth.
3. Humans have no right to reduce this richness and diversity except to satisfy *vital* needs.
4. The flourishing of human life and cultures is compatible with a substantial decrease of the human population. The flourishing of non-human life requires such a decrease.
5. Present human interference with the non-human world is excessive, and the situation is rapidly worsening.
6. In view of the foregoing points, policies must be changed. The changes in policies affect basic economic, technological, and ideological structures. The resulting state of affairs will be deeply different from the present and make possible a more joyful experience of the connectedness of all things.
7. The ideological change is mainly that of appreciating life quality (dwelling in situations of inherent value) rather than adhering to an increasingly higher standard of living. There will be a profound awareness of the difference between big and great.
8. Those who subscribe to the foregoing points have an obligation directly or indirectly to participate in the attempt to implement the necessary changes.

Though a full exposition would be impossible here, each point of the platform deserves some clarification within the context of music education. Point 1 expresses the core deep ecological premise, that of inherent (intrinsic) value. It is the primary challenge of deep ecology that will be explored in this chapter. Point 2 seems to parallel scholarship in ethnomusicology and praxial music education that finds value in understanding and performing various musics. "All musics have social values, or they wouldn't exist" (Regelski 2016: 111). The deep ecological focus on diversity contributing to flourishing might enrich some educators' approach to spiritual values, "constructed within the frameworks of particular social and cultural contexts rather than universally constructed and mandated" (Jorgensen 2003: 28–29). Diversity can become a call to preserve various musical traditions, and, inspired by deep ecology, to extend that preservation to the musicking of non-human beings. This is especially important during the sixth great extinction.

Point 3 proposes the concept of *vital needs*. This was conceptualized as "deliberately vague to allow for considerable latitude in judgment" (Naess 2005: 69). Many music educators have a lot of autonomy when it comes to classroom purchases. A music educator should apply *vital needs* when determining whether to purchase a new technology for his/her classroom. New devices present benefits and challenges for the eco-literate pedagogue, because educational institutions produce too much waste (an ecological crisis). Based on observation of the *vital needs* of the students, the amount and distinctiveness of use a new device has an opportunity to facilitate and the likely longevity should be determined by the teacher, who best knows his/her students, school, and community needs. As Drengson (1995) observed, "supporters of the deep ecology movement humbly recognize human ignorance, and the need for cautious development of technology" (76). To exemplify, teachers who are unlikely to use a wide range of applications integrated with a diverse approach to pedagogy should not purchase iPads for their students, especially since iPads have such a limited longevity (new models are released every year), few are recycled, and those that are recycled often present health hazards to people in the two-thirds world where recycling occurs. But other teachers might make extensive use of the same tool to contribute to students' *vital needs*.

Point 4 deals with human population, which has little application to eco-literate music pedagogy. Broadly though, point 4 envisions a reconceptualization of families. Increased ecological awareness—even only becoming conscious of how much land is required to feed a family—can lead to more simple living and smaller families. Point 5 recognizes the challenges facing humanity, including music teachers and learners. Devall and Sessions (1985) recognize, "Excessive human intervention in natural processes has led other species to near-extinction. For deep ecologists the balance has long since been tipped in favor of humans. Now we must shift the balance back to protect the habitat of other species" (127). Music educators who recognize this challenge have a responsibility to cultivate an appreciation for wilderness to the fullest extent.

Music is able to be central to "joyful experience" that recognizes the interconnectedness of human life, non-human life, and nature as recommended in point 6. Interconnectedness is valued in Satis Coleman's 1920s–1940s music education philosophy (Shevock 2015). Through musicking we recognize relationships, and can change the values that underlie our political and other social actions (Small 1998). Musicking with recorded animal songs, perhaps modeling Paul Winter's earth music, in the classroom can provide a way to change the way we value non-human musics and lives. According to deep ecology's best-known poet, Gary Snyder (1995), "Nature happens, culture happens, *somewhere*" (240, emphasis in original). Students can improvise with recordings of whales, or wolves, or birds and can cultivate their musicianship concurrent with their sonic place-consciousness and increase how much they value local non-human musickers.

For music educators, point 7 seems connected to points 3 and 5. When we take students outside to music, we cultivate "dwelling in situations of inherent value," rather than institutionalized, human-made spaces. Also, when we choose quality over quantity, instead of adopting every technology we cultivate a more sophisticated view of technology. Climate change activist Bill McKibben (2003) identified this type of sophistication in *Enough*, wherein he uses the Amish practice of adopting or refusing new technologies based on a firm set of values, and how they will affect the quality of their community living, as a "sophisticated" approach to technology. Without necessarily accepting Amish values, music educators can learn from this sophisticated approach, and develop a set of rules that guide when to adopt and when to say "enough."

Point 8 acknowledges that music educators who recognize the intrinsic value of non-human musics and life have a responsibility to act. This is why deep ecology has been recognized as a radical and activist philosophy. Earth First![6] has been the most visibly deep ecology influenced organization, but other organizations (including the Sierra Club and 350.org) have, through various actions, acted in a way consistent with deep ecology.

Naess (2005) says, "The platform of the deep ecology movement can be grounded in religion or philosophy, including ethics. It can be said to be *derived* from these fundaments" (62, emphasis in original). Because the platform is a platform, and not a root cause (or "fundament"), in the next part of this chapter I will use two stories to construct my fundamental reasoning for adopting the platform and a generalized deep ecological stance. This argument is rooted in a certain type of Catholicism (Illich drew from Christian anarchist scholarship), and the ecofeminist view that the unjust treatment of women parallels the unjust domination of the environment in patriarchal societies. Thus this argument represents *a* root for an ecocentric valuing in music education and not *the* root for all people in all places. In other words, this provides a model for constructing your own music education on soil. The spiritual facets of this root will be clarified in Chapter 5, "Spiritual Praxis."

The Good Samaritan

"Story is a means of sense making, a way in and through which we represent, interrogate, and interpret experience and come to know ourselves and others" (Barrett and Stauffer 2012: 1). A familiar story might allow its message to go unnoticed. Bowers (2000) reflects on this in his discussion of cultural root metaphors:

> The root metaphors of a cultural group are based on its ancient mythopoetic narratives (or cosmology). . . . A root metaphor is distinctive in that it forms the basis of thinking and acting in many areas of cultural life. . . . Root metaphors provide conceptual and moral coherence to a culture. They often go unnoticed because they are reproduced through linguistic processes that are mostly taken for granted. (26–27)

"The Good Samaritan" exemplifies this type of familiarity. It is a story I grew up with. When I conceive of caring for people, this story comes to mind. But familiarity "disguises the shocking character" (Illich and Cayley 2005: 50) of the story. This story is shared to provide insight into a narrative central to education, a caring institution.

> [A scholar of the law] said to Jesus, "And who is my neighbor?" Jesus replied, "A man fell victim to robbers as he went down from Jerusalem to Jericho. They stripped and beat him and went off leaving him half-dead. A priest happened to be going down that road, but when he saw him, he passed by on the opposite side. Likewise a Levite came to the place, and when he saw him, he passed by on the opposite side. But a Samaritan traveler who came upon him was moved with compassion at the sight. He approached the victim, poured oil and wine over his wounds and bandaged them. Then he lifted him up on his own animal, took him to an inn and cared for him. The next day he took out two silver coins and gave them to the innkeeper with the instruction, 'Take care of him. If you spend more than what I have given you, I shall repay you on my way back.' Which of these three, in your opinion, was neighbor to the robbers' victim?"
>
> (Lk 10:29–37 NABRE)

After conducting a survey of sermons on "The Good Samaritan," Illich concluded that most understand this story as illuminating what duty one has toward one's neighbor—how one should act (Illich and Cayley 2005: 50–53). Directing our ears to the story's opening question—"And who is my neighbor?"—Illich reveals that this story represents, rather, an expansion of ethics beyond the *ethnos* (47): "outside his ethnic preference for taking care of his own kind" (51). By expanding ethical action beyond ethnic (*ethnos*) boundaries, "The Good Samaritan" represents a mythopoetic narrative of who is able to care for whom. However, when we institutionalize this trans-*ethnos* care, we take away the ability for the Samaritan to choose to act—care is done by hospitals, schools, and other institutions. Illich alleged, "Wherever I look for the roots of modernity, I find them in the attempts of the churches to institutionalize, legitimize, and manage Christian vocation" (48). Over centuries Western society institutionalized the then new trans-*ethnos* values into the caring professions. As a result, Western education has as its aim the educational care for people beyond ethnic boundaries (what we often call "inclusion"). The challenges inherent in this aim have sustained much music education scholarship. Illich's work established the historical extension of ethics from within-*ethnos* to trans-*ethnos* and the institutionalization of values in the establishment of caring professions. By telling a second story, synchronized with "The Good Samaritan," the advantage to extending trans-*ethnos*, human-centered (anthropocentric) values to ecocentric valuing in music education is explored. I hope that music education on soil, being a rooted, place-conscious practice, might avoid many of the negatives Illich connected to the institutionalization of values.

Creative Vignette

The small dam was old and unproductive. It never produced as much electricity as promised, and lately the sediment created a minor ecological crisis. The town's water had needed to be boiled for five years now. Once steelhead trout flourished in this creek. The oldest grandparents in town had cherished memories of pulling them from the waters. Reading how dam removal had helped to revitalize trout populations in other creeks and rivers, the grandparents felt that this might be the best option, and a community meeting was called to explore dam removal. With little economic incentive, the local government rejected the grandparents' request.

Meeting weekly, the grandparents decided to organize a festival, Free the Creek, to ritualize[7] their desire for dam removal. They began

(continued)

(continued)

by talking about their youthful memories: not only the good times fishing, but also the games they played along the creek, the rhymes and songs they sang, the dishes they ate, and the old stories they used to tell. This festival, which would be held on a field near the dam, would celebrate the local place. From these memories, they produced a songbook for the culminating sing-along, which would, by the glow of a bonfire and the full moon, feed hope for the townsfolk struggling to reclaim their home. They hoped to raise funds to assist with dam removal, but, more importantly to them, they hoped to help to reconnect young people (who too often leave for forgettable suburbia to participate in the global economy) to the beauty of this soil, their place.

The Birth of Pecos Bill

> Well now Pecos Bill was born in the usual way to a real nice cowpoke and his wife who were journeying west with their eighteen children. . . . When he started to crawl, Pecos Bill would slither out of the wagon while his Mama was cookin' supper and wrestle with the bear cubs and other wild animals that roamed the prairies. Yep, the whole family was expecting great things of little Bill; until they lost him in . . . the Pecos River. If he hadn't taught himself to swim right-quick, he would have been a goner!
>
> Right about the time Pecos Bill was drying out and trying to get a fix on where he was, a Mama Coyote came along and decided to adopt the poor waif and raise him with the rest of her pups. So Pecos Bill spent the first fifteen years of his life running around with the coyote pack, howling to the moon, chasing prey across the prairies, and having the time of his life.
>
> (Schlosser n.d.)

Stories of wild animals raising young children abound in fiction and real life. In myth, famous examples include Romulus and Remus, Mowgli (*The Jungle Book*), Tarzan, and Enkidu (*Epic of Gilgamesh*). I chose a mythological example to explore the parallels between Mama Coyote, who rescued and cared for Pecos Bill, and "The Good Samaritan." If, as Illich suggested, "The Good Samaritan" can be understood as an expansion of ethics beyond the *ethnos* (Illich and Cayley 2005: 47), "The Birth of Pecos Bill" might be understood as an expansion of ethics beyond the anthropocentric. If the Samaritan represents a choice to act ethically outside of an ethnic group, the Mama Coyote represents a choice to act ethically beyond her species. It is also interesting for music educators that the coyote intones a diverse range of sounds and survives in wilderness and near cities throughout North America. If we can say our familiar sister coyote deserves a parallel ethical consideration, can we also attribute intrinsic value to her musicking? In story at least, she has taken the first step in offering humans ethical consideration.

Social Foundations of Valuing in Music Education

Understanding music education as social praxis helps teachers and students come to understand and value human relationships, and shapes how we understand those relationships. Music is a social act (Elliott and Silverman 2015; Goble 2010; Regelski 2016; Small 1998). Small clarified, "There is no such thing as music. Music is not a thing at all but an activity, something that people do" (2). For Small, musicking forms stories that support a social order—that is, how humans relate to one another (89–93). For music educators involved in teaching music as social value, understanding the social value of your musics and the musics of *the other* becomes important. Goble (2010) concludes music educators should "come to engage with [other] musical practices and understand them *on their own terms*, thereby gaining intercultural knowledge and potentially developing greater interpersonal understanding" (279, emphasis in original).

Elliott and Silverman's (2015) praxial philosophy for music education is founded on the social value of music. They outline ways of conceiving a specific musical praxis: 1) the "person(s), or 'musical doers'"; 2) the "musical processes or 'doings'"; 3) the "musical products or 'outcomes'"; and 4) the "musical contexts" (210–211). To expand upon musical contexts, they state, "by this we mean all possible musical spaces and places" (211), including "World-Environmental Contexts" (210). For deep ecologists, environmental contexts encompass all social contexts. "Human beings, biology and ecology tell us, are located completely within the sphere of nature. Social organization, language, cultural practices, and other features that we take to be distinguishing characteristics of the human species are also within the larger sphere of nature" (Snyder 1995: 237). Social contexts are within nature.

Rootedness is also important to the social foundations of music education. Elliott and Silverman (2015) write, "The musical understanding underlying any praxis of music making and listening has its root in specific communities of practitioners who share and advance specific forms of musical thinking and doing" (228). In eco-literate music pedagogy, this rootedness in *specific communities of practitioners* is extended (by truncating the phrase) to *specific communities*. This allows us to focus in on roots, that is, actual people in actual places living and musicking together. Consistent with Small's (1998) inclusive descriptions of who musics, music educators are forming pedagogies rooted in communities that shape who students are and respect diversity (e.g., Lind and McKoy 2016: 116–130). Rootedness is inextricably connected to place (a plant's roots touch specific, not conceptualized, soil), which is historically understood by people who live there (often through storytelling) and linked to local ways of understanding how to live sustainably (see Bates 2011; Prakash and Esteva 2008; Shevock 2016; Weil 1952). Because of this, deep ecology allows us to extend praxialism by looking downward, to soil where virtue is cultivated (see Chapter 1, "Philosophy on Soil"). The deep ecological praxial arrangement of music within place and nature compels me to appraise musicking within the context of lived-in soundscapes.

Soundscapes

Schafer (1994) drew our attention to the soundscape—to see that the soundscape is changing negatively for modern people (3–4). He also drew our attention to the musical sounds of life (29–42), including the analysis of "the songs of humpback whale[s] . . . in musical terms" (37), and the sounds of animals, including the "haunting and isolating" (39) howling of wolves. Schafer's soundscape is difficult to label as either anthropocentric or ecocentric. For instance, his description of wolf howling as "haunting and isolating" [to humans] seems to point toward anthropocentric valuing. When discussing birdsongs, Schafer suggested, "despite . . . similarities [between human and bird expressions], it is obvious that to whatever extent the birds are deliberately communicating, it is for their own benefit rather than ours that their vocalizations are designed. . . . Birds, like poems, should not mean, but be" (31). This sentence seems to represent the paradox inherent in Schafer's conception of soundscape—birds music *for their own benefit* (ecocentric), but can be compared to poems, which are human-made (privileging the human-made is anthropocentric). For some music educators, it may be important to let this paradox lie as it is, as it provides insight into the deep ecological understanding that humans are within the larger sphere of nature, but that non-human musics are made sense of by humans as social and symbol-constructing beings. Are we touching non-human life when we use non-human musics in our own improvisations and compositions? We may be cultivating our capacity for empathy.

Parallels of Injustice

Following Goble's (2010) reasoning, in gaining intercultural knowledge through music we gain insight into the ways people live together. The ways we live together can be either positive or negative. Ecofeminists[8] have theorized how the "unjustified domination of women" parallels the "unjustified domination of non-human nature" (Warren 2000: xiv). Warren reveals historical, conceptual, linguistic, symbolic and literary, spiritual and religious, epistemological, political, and ethical interconnections between these unjustified dominations (21–39). She uses the terms "Up" and "Down" to illustrate. Oppressive conceptual frameworks sanction:

> a *logic of domination*, that is, a logical structure of argumentation that "justifies" domination and subordination. . . . Typically this justification takes the form that the Up has some characteristic (e.g., in the Western philosophical tradition, the favored trait is "mind," reason, or rationality) that the Down lacks and by virtue of which the subordination of the Down by the Up is justified. (47)

In this *logic of domination*, music educators might choose repertoire by White men, assuming this music is more mindful, reasoning, and rational. Further, the musics of nature may be excluded entirely because it is not mindful, reasoning, and rational, and we strengthen the unjust domination of nature.

Ecofeminism can also be a way to deepen the criticism of anthropocentric valuing (Warren 2000: 24). One way to criticize anthropocentric valuing is to attribute intrinsic value to non-human life. Ecofeminist Vandana Shiva (2005) acknowledges: "Living democracies are based on the intrinsic worth of all species, all peoples, all cultures" (6). Connecting social and ecological concerns makes possible the extension of intrinsic value to non-humans.

Musics have social value, and this value is intrinsic to the people who are musicking. Because "humans must learn to live in balance with the needs of nature" (Bari 1994: 220), musics might reinforce values of domination or living democracies (Shiva 2005). When considered along with Schafer's (1994) birds, who music "for their own benefit rather than ours" (31), Goble's (2010) argument for music education fostering intercultural knowledge can be extended to include music educators engaging with the musics of non-human life, and developing a greater diversity of musical understandings than can be developed by engaging with human cultures alone. If musics of non-human life are understood as having intrinsic value (intrinsic to those life forms who are musicking), music education can cultivate interspecies knowledge.

A Valuing Structure

The valuing structure I propose draws on both deep ecology and ecofeminism. It can be challenging to move music teachers and students beyond the current anthropocentric valuing—part of the patriarchy that unjustly dominates women and nature—to ecocentric valuing. Conscientization can be the heartbeat of cultivating ecocentric valuing in music education.

1 Conscientization can begin with becoming aware of our society's taken-for-granted values (including stories and musics) that underpin the unjust domination of nature.
2 Conscientization can be directed toward ecological ends by recognizing the intrinsic value of non-human musicking (life forms musicking *for their own benefit*).
3 As we become conscious of the ways we unjustly dominate and the intrinsic value of non-human musicking, we can begin to realize solidarity with nature, moving beyond the anthropocentric.

Ecocentric valuing of non-human life and musicking might inspire activism as we address our shared ecological crises. This valuing structure does not provide prescriptive answers to the ecological crises—climate change, loss of species diversity, destruction of wilderness, plastic waste, collapse of fish populations, soil degradation, lack of apex predators, and e-waste to name a few. Instead, this valuing system provides a foundation for music teachers and students to begin the process of analyzing these ecological crises beginning with the possible distinct ways of thinking that musics afford.

In this chapter I drew parallels between two narratives—"The Good Samaritan" and "The Birth of Pecos Bill." The two main characters, the Samaritan and Mama Coyote, provide narrative ways of understanding an extension of values (and hospitality, and care): first beyond ethnic groups and then beyond species. The social value of music education, ecofeminist philosophy, and deep ecology provided a basis for ecocentric valuing in music education. I suspect that, paralleling deep ecology, ecocentric valuing in music education might provide inspiration for both scholarship and activism.

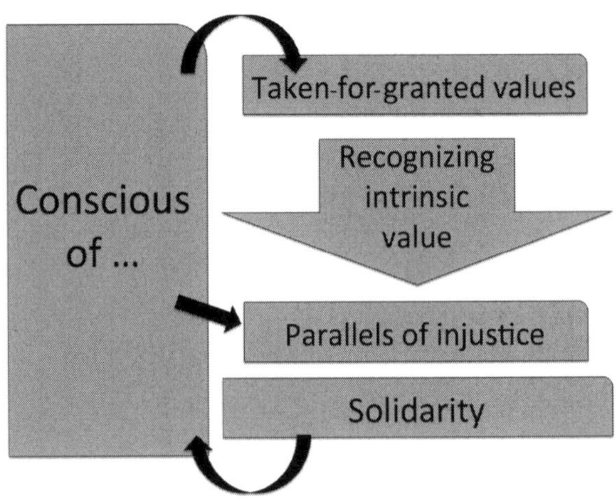

Figure 4.1 A Valuing Structure

Challenging the Metaphor

Because deep ecology discusses non-human life and nature in terms of "rights," deep ecology provides a metaphor about nature that is both useful and problematic. In his chapter on planting a tree in his backyard, Michael Pollan (1991) discusses a variety of metaphors for how humans think of trees: the Political Tree, the planting of which is "a patriotic act" (163), and which provides wood for human use and delineates property borders; the Romantic Tree with its spiritual significance (Thoreau loved this metaphor); the Litigious Tree which possesses rights, a metaphor that might be most consistent with the deep ecology movement; the Lung Tree (science has described trees as the lungs of our larger ecosystem); the Canary Tree, which, like the canary in the mineshaft, warns us of oncoming ecological disaster just in time; but that what we really need is "evidence of a new metaphor, something vivid and powerful and, for a time at least, *true*" (175, emphasis in original). If deep ecology did not contain a spiritual aspect, the challenge of its central metaphor (a metaphor that itself challenges the "resource" metaphor) might be even more damning than it is. Spiritual practices can accept, and even celebrate, paradoxes.

From a postmodern position, each of Pollan's metaphors (and the search for a single "new metaphor") reveals a grand narrative, which is problematic (see Lyotard 1984). While construction of metaphors is a human activity—perhaps even anthropocentric—humans using all of these metaphors can better understand nature. Thomas Berry's (1999) nature as a symphony[9] metaphor provides insight into nature. But why only a symphony? Might nature also be understood as a country ballad (expressing unfulfilled love or nostalgia), or politically charged rap (sharing hardships that the dominant culture doesn't hear), or free jazz (with emerging, complex, unpredictable forms, rhythms, harmonies, and melodies)? This might help us move from the understandable to the possible (see Allsup 2016: 132–135). Each of these metaphors serves as a temporary construct, and each layers upon the other—topsoil upon subsoil upon parent rock upon bedrock—each gifting new and old possibilities.

Even so, Berry, as a Catholic ecotheologian, recognized that changing metaphors might not be enough. Metaphors are part of rationality, and I have used them in this chapter to construct a root cause for ecocentric valuing in this philosophy of music education on soil. Dolores LaChapelle (1995) suggested, "If we are to truly re-connect with the land, we need to change our perceptions and approach more than our location. As long as we limit ourselves to rationality and its limited sense of 'practicality,' we will be disconnected from the 'deep ecology' of our place" (224). Deep ecologists have recommended *ritual* as a way to connect, as part of nature, to our ecologies (e.g., Devall and Sessions 1985; LaChapelle 1995). In the next chapter, I will extend eco-literate music pedagogy into the realm of the spiritual, and advance music education on soil as a spiritual praxis of music teaching and learning.

Birdsong

birdsong
symphonizes this morning:
for whose benefit?
crows caw
and sparrows chip:
for whose benefit?
and spontaneously
—freely
children dance and sing:
for whose benefit?

Notes

1 From *Earth Democracy* (Shiva 2005: 2), which draws together ecofeminist and deep ecological thought.
2 According to Warwick Fox (2000), deep ecology (a so-called radical ecology, along with ecofeminism and social ecology) is based on the premise of ecocentrism, that is, challenging anthropocentric (human-centered) valuing and acting. Both deep ecologists and ecofeminists (such as Shiva 2005) often draw on ecocentrism.
3 In music education scholarship, there has been scholarship around the idea of intrinsic value in music—that is, that music (the sounds) has some sort of internal value (e.g., a value music might have on Alpha Centauri without humans present). The distinction made in this literature is between intrinsic and extrinsic valuing in music. This distinction is not helpful for our current chapter, because it fails to address how we should act toward other people and nature; and it fails to address how music helps us relate ethically to other people and nature.
4 Although it seems that pre-civilized (civil = cities) cultures were, generally speaking, ecocentric, anthropocentrism has been the dominant philosophy in Western scholarship. According to Sessions (1995), throughout the history of the West, figures as diverse as Francis of Assisi, Baruch Spinoza, Jean-Jacques Rousseau, John Stuart Mill, Charles Darwin, Henry David Thoreau, George Perkins Marsh, John Muir, and George Santayana have put forth non-anthropocentric theories. However, these thinkers were unable to persuade the West into "leav[ing] the path of the ecologically destructive 'anthropocentric detour'" (160). It was "Lynn White's critique of Christian anthropocentricism and the ecological crisis, together with Aldo Leopold's ecocentric perspective, [which] were major influences on the rise of ecophilosophy, as well as ecotheology, in the late 1960s and 1970s" (156). Wolfgang Sachs (1995) connected anthropocentricism to development discourse and environmental degradation: "With 'development' back in the saddle, the view on nature changes. The question now becomes: which of nature's 'services' are to what extent indispensable for further development" (434, emphasis in original). With development discourse's prominence today (e.g., sustainable development, underdeveloped nations, childhood development, etc.), ecocentrism becomes crucial for educators hoping to facilitate students' understanding of root causes of the ecological crises.
5 Link: http://www.deepecology.org/platform.htm.
6 This organization has been feminized through ecofeminist deep ecologists like Judi Bari (1994), who stated, "Little did I know that by combining the more feminine elements of collectivism and non-violence with the spunk and outrage of Earth First!, we would spark a mass movement" (219).
7 Deep ecology often contains a spiritual element that can be expressed through many religious traditions.

> And now the sky and water bear our mark, and we seek to make ourselves masters of birth and death, [spiritual deep ecology] has returned. Spiritual deep ecology faces a divinity that is itself under siege. We are called not only to obey the teachings of the more than human, but to save it from ourselves.
> (Gottlieb 2001: 31).

Because of this ecumenical-ecological valuing, deep ecologists ritualize places in a consciousness-raising spirit of interconnectedness.

8 To the best of my knowledge, Roberta Lamb (2014) is the only music education scholar to draw from ecofeminist literature. Her paper points to the irrationality of both social and environmental actions and to "another possible world" (213).
9 This metaphor was shared in Chapter 3, "Ecomusicology," because it relates musical thinking to understanding the ecological crises:

> We need to see the Earth in its sequence of transformations as so many movements in a musical composition. The sequence of events that emerge in time needs to be understood simultaneously, as in music: the earlier notes are gone when the later notes are played, but the musical phrase, indeed the entire symphony, needs to be heard simultaneously. . . . Each new theme alters the meaning of the earlier themes and the entire composition. The opening theme resonates throughout all the later parts of the piece.
>
> (Berry 1999: 27).

References

Allsup, Randall Everett. 2016. *Remixing the Classroom: Toward an Open Philosophy of Music Education*. Bloomington: Indiana University Press.

Bari, Judi. 1994. *Timber Wars*. Monroe, ME: Common Courage Press.

Barrett, Margaret S., and Sandra L. Stauffer. 2012. "Resonant Work: Toward an Ethic of Narrative Research." In *Narrative Soundings: An Anthology of Narrative Inquiry in Music Education*, edited by Margaret S. Barrett and Sandra L. Stauffer, 1–22. New York: Springer.

Bates, Vincent C. 2011. "Sustainable School Music for Poor, White, Rural Students." *Action, Criticism, and Theory for Music Education* 10, no. 2: 100–127.

Berry, Thomas. 1995. "The Viable Human." In *Deep Ecology for the 21st Century: Readings on the Philosophy and Practice of the New Environmentalism*, edited by George Sessions, 8–18. Boston, MA: Shambhala Publications.

——. 1999. *The Great Work: Our Way into the Future*. New York: Bell Tower.

Bowers, C. A. 2000. *Let Them Eat Data: How Computers Affect Education, Cultural Diversity, and the Prospects of Ecological Sustainability*. Athens: University of Georgia Press.

——. 2006. *Revitalizing the Commons: Cultural and Educational Sites of Resistance and Affirmation*. Lanham, MD: Lexington Books.

Devall, Bill, and George Sessions. 1985. *Deep Ecology: Living as if Nature Mattered*. Layton, UT: Gibbs M. Smith.

Drengson, Alan. 1995. "Shifting Paradigms: From Technocrat to Planetary Person." In *The Deep Ecology Movement: An Introductory Anthology*, edited by Alan Drengson and Yuichi Inoue, 74–100. Berkeley, CA: North Atlantic Books.

Elliott, David J., and Marissa Silverman. 2015. *Music Matters: A Philosophy of Music Education*, 2nd edition. New York: Oxford University Press.

Fox, Warwick. 1995. *Toward a Transpersonal Ecology: Developing New Foundations for Environmentalism*. Albany: State University of New York Press.

——. 2000. "Deep Ecology and Virtue Ethics." *Philosophy Now* 26 (April/May): 21–23.

Freire, Paulo. 2000. *Pedagogy of the Oppressed*. New York: Bloomsbury Publishing. [Originally published in 1970.]

Goble, J. Scott. 2010. *What's So Important about Music Education?* New York: Routledge.

Gottlieb, Roger S. 2001. "Spiritual Deep Ecology and World Religions: A Shared Fate, a Shared Task." In *Deep Ecology and World Religions: New Essays on Sacred Ground*, edited by David Landis Barnhill and Roger S. Gottlieb, 17–33. Albany: State University of New York Press.

Illich, Ivan and David Cayley. 2005. *The Rivers North of the Future: The Testament of Ivan Illich as Told to David Cayley*. Toronto: House of Anansi Press.

Jorgensen, Estelle R. 2003. *Transforming Music Education*. Bloomington: Indiana University Press.

LaChapelle, Dolores. 1995. "Ritual Is Essential." In *The Deep Ecology Movement: An Introductory Anthology*, edited by Alan Drengson and Yuichi Inoue, 219–225. Berkeley, CA: North Atlantic Books.

Lamb, Roberta. 2014. "Where Are the Women? And Other Questions, Asked within an Historical Analysis of Sociology of Music Education Research Publications: Being a Self-Reflective Ethnographic Path." *Action, Criticism, and Theory for Music Education* 13, no. 1 (March): 188–222.

Lind, Vicki R., and Constance L. McKoy. 2016. *Culturally Responsive Teaching in Music Education: From Understanding to Application*. New York: Routledge.

Lyotard, Jean-François. 1984. *The Postmodern Condition: A Report on Knowledge*. Minneapolis: University of Minnesota Press.

McKibben, Bill. 2003. *Enough: Staying Human in an Engineered Age*. New York: Henry Holt and Company.

Naess, Arne. 2005. "The Basics of Deep Ecology." *Trumpeter* 21, no. 1: 61–71. Revised version of a paper given in Canberra, Australia in 1986.

Pollan, Michael. 1991. *Second Nature: A Gardener's Education*. New York: Atlantic Monthly Press.

Prakash, Madhu Suri, and Gustavo Esteva. 2008. *Escaping Education: Living as Learning within Grassroots Cultures*, 2nd edition. New York: Peter Lang.

Regelski, Thomas A. 2016. *A Brief Introduction to a Philosophy of Music and Music Education as Social Praxis*. New York: Routledge.

Rehding, Alexander. 2002. "Review: Eco-Musicology." *Journal of the Royal Musical Association* 127, no. 2: 305–320.

Sachs, Wolfgang. 1995. "Global Ecology and the Shadow of 'Development.'" In *Deep Ecology for the 21st Century*, edited by George Sessions, 428–444. Boston, MA: Shambhala Publications.

Schafer, R. Murray. 1994. *The Soundscape: Our Sonic Environment and the Turning of the World*. Rochester, VT: Destiny Books. [Originally published in 1977.]

Schlosser, S. E. n.d. "The Birth of Pecos Bill: A Texas Tale." Online: http://americanfolklore.net/folklore/2010/08/the_birth_of_pecos_bill.html.

Sessions, George. 1995. "Ecocentrism and the Anthropocentric Detour." In *Deep Ecology for the 21st Century*, edited by George Sessions, 156–183. Boston, MA: Shambhala Publications.

Shevock, Daniel J. 2015. "Satis Coleman—A Spiritual Philosophy for Music Education." *Music Educators Journal* 102, no. 1 (September): 56–61. DOI: 10.1177/0027432115590182.

——. 2016. "Music Educated and Uprooted: My Story of Rurality, Whiteness, Musicing, and Teaching." *Action, Criticism, and Theory for Music Education* 15, no. 4 (August): 30–55. DOI: 10.22176/act15.4.30.
Shiva, Vandana. 2005. *Earth Democracy: Justice, Sustainability and Peace*. Cambridge, MA: South End Press.
Small, Christopher. 1998. *Musicking: The Meanings of Performing and Listening*. Middletown, CT: Wesleyan University Press.
Snyder, Gary. 1995. "Ecology, Place, and the Awakening of Compassion." In *The Deep Ecology Movement: An Introductory Anthology*, edited by Alan Drengson and Yuichi Inoue, 237–241. Berkeley, CA: North Atlantic Books.
Warren, Karen J. 2000. *Ecofeminist Philosophy: A Western Perspective on What It Is and Why It Matters*. Lanham, MD: Rowman & Littlefield Publishers.
Weil, Simone. 1952. *The Need for Roots*. New York: Routledge.

5 Spiritual Praxis

> Today, however, we have to realize that a true ecological approach always becomes a social approach: it must integrate questions of justice in debates on the environment, so as to hear *both the cry of the poor and the cry of the earth.*
>
> (Pope Francis, 2015[1])

Three Glimpses

The morning sun glimmers through the red and blue stained glass as the congregation sings "For the Beauty of the Earth." In communion with God and neighbors, I want to be a better steward for God's environment.

Outside, under the chilly Beltane sun, a drummer plays and we sing a druid elder's song, "We Are One Family." In this community, dancing with friends and strangers, I become kin to non-human animals and plants.

I sit beneath an old maple and meditate on the Zen koan "A Drop of Water." What right do I have to waste even a single drop?

On Earth as It Is in Heaven

During first period prep, when the towers fell that disastrous September morning, New York didn't feel distant. The Pentagon didn't feel distant. Told not to tell the students, the principal sent hand-written updates to us throughout the day. Another plane followed, falling outside of Pittsburgh, where I taught elementary general music. Leading the children in song was difficult for me. But they were unaware and sang strongly through second period. By fourth period the fifth grade class—the half still in class—knew something was wrong. Why had so many of their classmates been picked up early?

In loco parentis—the phrase had never meant anything to me before. Now it weighed heavily. Only twenty-six years old. Not a parent. Who was I to be responsible for these children? And how horrible would today become before it ended? What can I do? Would planes continue to fly into buildings? How long could I . . . should I . . . keep my students unaware? They saw teachers crying. They heard hushed voices.

I didn't lose my faith on 9/11/2001. Hyper-nationalistic, conservative, anti-Muslim rhetoric following shook my faith in America and in the Catholic Church. I read and prayed, unable to make sense of this senseless day. I entered what St. John of the Cross named *the dark night of the soul*. Not finding answers in an ostensibly rightwing Church, finding tradition insufficient, and finding my prayers unanswered, I put my ear to the ground. The first prayer I learned was the Our Father. Taking those words to heart, "on earth as it is in heaven," I began to look below my feet.

On the earth I disavowed solace I once took in tradition, dogma, and doctrine. I stopped going to church and started going to the woods. I allowed Mother Earth to become my parent. Studying Buddhism and various earth religions, I found sustenance in being with communities connected to each other and nature. I searched in these other traditions for The Truth for years. I met many friends who were also searching, but could not unearth The Truth. Without finding it, I settled for gaining some wisdom along the way. My path, which is mine and mine alone (a paradox), led back to the Catholic Church. Jesus, who meditated in the wilderness before beginning his mission, preached the Beatitudes and the Prodigal Son, and gave us the Our Father, called me to return. Revisiting authors I had not in years, I rediscovered the profoundly life-giving, ecologically sustainable, and deep-rooted spirituality of my youth, my family, and my church. This rediscovery was new. I had been through my *dark night of the soul*, and discovered a more interconnected and religious—"reconnected [*re-ligado*] to the divine heart" (Boff 2001: 33)—spirituality.

Spiritual Aims

Contemplating the ecological crises is disheartening. Centuries of unrelenting environmental degradation continue, and two of our society's trusted remedies—government policy and technological development—seem feeble. Some policies have been successful, especially decades-old policies such as in the U.S. the 1963 Clean Air Act, 1972 Clean Water Act, 1980 Comprehensive Environmental Response, Compensation, and Liability Act (CERCLA), and 1973 Endangered Species Act. However, international action on climate change has been less effective. The 1992 United Nations Framework Convention on Climate Change treaty in Rio de Janeiro famously identified a wish to halt climate change at a 2 degree Celsius increase, "to 'avoid dangerous anthropogenic climate change.'"[2] However, the changes each nation would have to legislate to meet the 2 degree Celsius goal were identified as "voluntary," and those changes have not been implemented. In the current chapter, music education on soil, as rooted and connecting, is understood as spiritual praxis cultivating the hope we need to face the ecological crises.

I work for a university, and universities love technological fixes. These should be explored. Solar, geothermal, biofuel, and wind technologies have made advances. But big oil industries began pulling out of solar, geothermal, and wind investments in 2009 (and some companies, like Exxon Mobil, have never invested in sustainable energy). This coincides with increases in highly polluting alternative extraction techniques such as fracking, mountaintop removal, and deep water and Arctic drilling. Natural systems scientist Stephen Emmott (2013) concludes, "All the science points to the inescapable fact that we are in trouble. Serious trouble" (129), and, "on today's evidence, technologizing our way out of this does not look likely" (184). With such a negative forecast, can we do anything, now, even while policy and technology seem ineffective? Hiding our heads in the sand isn't a viable option.

Wendell Berry (2010) appropriately notes the ecological challenges are not political, but cultural at root (37). Policy, then, without a parallel change in cultural beliefs, will be insufficient. In my philosophy of music education on soil (outlined in Chapter 1, "Philosophy on Soil"), the cultural disconnection that led to our ecological crises is understood as a spiritual problem, and requires a spiritual reconnection to neighbors, to actual places, and to an expanded view of self that values non-human life forms and their musicking intrinsically (Chapter 4, "Deep Ecology"). Using a phenomenological sentence structure (Husserl 1983), I can write:

Sentence 1: Musicker musics for spiritual aims.[3]

This sentence will be theoretically expanded throughout this chapter, and labeled "Sentence [number]."

We begin to heal the earth by recognizing our participation in the ecological crises. Henri Nouwen writes, "In the solitude of the heart . . . we can feel that the cruel reality of history is the reality of the human heart, our own included, and that to protest asks, first of all, for a confession of our own participation in the human condition" (Jonas 2009: 176). The underlying solitude of the heart seems to involve a mode of *having* rather than *being* (Fromm 1976).[4] Participating in industrial society, we built, sold, bought, spoiled, and threw away our way into ecological crises. This is the *having* mode. We have become disconnected[5] from one another, and we further exacerbate the having mode. Degrading ourselves, we become *human havings* rather than *human beings*. It should be no surprise *human havings* degrade the environment.

People need roots (Weil 1952), but our current education system educates people "to leave home. . . . One of the costs is psychological, and the other is at once cultural and ecological" (Berry 2010: 164). When we are disconnected from one another, we are unable to live and act well.[6] Feeling connected to people and places is psychologically beneficial. For instance, people who feel connected to nature experience reduced anxiety (Martyn and Brymer 2016). But education has historically been disconnecting. Feminist bell hooks (2003) suggests, "Conventional education teaches us that disconnection is organic to being" (180). A philosophy of music education on soil rejects disconnecting practices, because they are not organic or sustainable.

Scholars of many faith traditions have put forth ecotheological statements. Muslim scholar Seyyed Hossein Nasr (1990) suggests the problem is modern scientism, "a complete and totalitarian philosophy which reduces all reality to the physical domain and does not wish under any condition to accept the possibility of the existence of non-scientistic world-views" (4). This critique seems similar to critiques of scientism by Henri Bergson, Herbert Marcuse, and Erich Fromm, but from a religious point of view. Nasr continues, "the ecological crisis cannot be solved without paying particular attention to the spiritual dimension of the problem" (7). Living in harmony with nature requires looking to the source of nature. Because deep ecology involves "exploring ecological self . . . part of the transforming process required to heal ourselves in the world" (Devall 1988: 42), deep ecologists understand themselves as scientific, psychological, social, and spiritual beings.

In previous chapters, eco-literate music pedagogy was presented as music education on soil, rooted in place and respectful of tradition. Delving into interdisciplinarity and ecomusicology expanded understandings of eco-literate music pedagogy. Music education on soil was deepened to emphasize the intrinsic value of non-human being and musicking. In the current chapter, spirituality is theorized as praxial, guided by intentionality and understood as action, rather than only aesthetic. Ideas of Catholic and Buddhist ecotheologians and authors are used to justify a spiritual praxis for eco-literate music pedagogy. Living a good life involves an expanded understanding of self, interconnected to God, other humans, non-humans, and places.

The Possibility of Praxial Spirituality

Spirituality can be hard to define. Perhaps it should be hard to define—paradoxical even. Rather than define it, I am aiming for an understanding of spirituality as praxis. To the non-music educator, that music education is linked to spirituality and religion can seem obvious. Music is part of most religious activities, and has been since before written history. Satis Coleman (1928) speculates:

> In the early ages of the human race, when primitive man[7] first evoked a vibrant sound from a stone, it must have seemed to him the voice of his God speaking the mysterious language of Mother Earth. For was not stone a part of the sacred underworld? Was not the tree also attached to Mother Earth in a mysterious way, and would not the peculiar sound of hard wood, when struck, seem to have some hidden power attached to it?
>
> The strangeness of these sounds led primitive man to invest them with a sacred character. To his thinking, everything which he could not understand was something to be worshipped, and the sound of stricken metal and hollow wood must have conveyed to him much that was supernatural. His gods spoke to him in the crash of thunder, and in all other vibrant and mysterious sounds. (9)

Music and ritual are interconnected, and music educators can clarify those connections. In music education research on spirituality, interconnections go beyond connecting music and ritual, to "relationships between individuals and communities, the natural world, cosmos, and the divine" (van der Merwe and Habron 2015: 57).[8] Spirituality in music education can be understood as a drive to "explore the shared realm of human music experience" (Matsunobu 2011: 284).[9] Thus, to live the good life as a music teacher is, for many people, understanding spiritual aims. Jorgensen (2003) discusses spiritual values within the context of the Gaia hypothesis. This view emphasizes stewardship of nature, living in harmony with the environment, care, cooperation, recognizing interconnection with others, and emphasis on process over product (28–29). Spirituality, in music education, means looking for that which connects us to one another. In a philosophy of music education on soil, "us" is understood more broadly than only humans.

June Boyce-Tillman (2007) raises critical questions[10] when encouraging a spiritual praxis of music education: "Is the aesthetic/spiritual a contemporary version of the religious? \ Is there a notion of a spiritual dimension that can be freed from a particular religious tradition?" (1406). Though many music education scholars approach spiritual issues from the point of view of aesthetics (van der Merwe and Habron 2015), the current philosophy of music education on soil separates the aesthetic from the spiritual, because, in music education at least, aesthetics theorists seem to scientize (see *scientism* above) spirituality when spiritual aspects of musicking come under their purview. I recommend a spiritual praxis of music education. A praxial approach to doing spirituality may be unable, and may not desire, to separate *spirituality* from *religion*, and alternatively not-religion,[11] because the people intending spirituality through musicking have religious and/or non-religious aims. For some people spiritual actions are religious, and for others they're not. For a music teacher understanding his/her deepest motivations for choosing this ethical work, separating spirit and religion may be impractical. The way people intend musicking ritual can be an excellent way to illuminate spiritual praxis.

Some spirituality in music education scholars frame their work within the context of aesthetics. Aesthetic education focuses on perceiving works of art, often described using cognitive psychology and, secondarily, culture.

> Since man learns from experience, and since music is part of his experience, he has naturally developed (consciously or unconsciously) some degree of comprehension in musical matters. For the aesthetic experience to take place the respondent cannot simply bathe in the "emotional waters." The respondent is acutely conscious while affectively engaged. This intellectual awareness is usually manifested in one or more of the typical cognitive processes: analysis, synthesis, abstraction, generalization, evaluation.
>
> (Knieter 1971: 5)

Aesthetic educators aim to "cultivate aesthetic sensitivity" (6). However, the scientific activities—*analysis, synthesis, abstraction, generalization, evaluation*—approach scientism, reducing and qualifying non-scientific experiences (in this case, spirituality experienced by musickers) to scientific scrutiny. To attempt to understand spirituality (or any other cultural practice) by beginning with an individual's response to a work of art, and within the terminology of cognitive science, misses something important of what spirituality actually is. A Catholic nun meditating as she sings the *Kyrie*, or a Buddhist monk chanting to take refuge in the Three Jewels intends far more than mere cognitive benefits. An analysis of the *work of art* misses the point of the practice. The pitch relations, form, and other musical elements might be simple, but the intention complex. In this case, all of the critiques of music education as aesthetic education by praxialists are appropriate to the current philosophy of music education on soil. "Aesthetic rationales are simply not needed in order to account for the *moving power* of music and art" (Regelski 2016: 39).

Regelski (2016) goes further by suggesting aesthetic theorists value what I refer to as spiritual music less, ignoring the social praxis of these musics.

> Because music is held [by aesthetic theorists] to be autonomous and "in" the mind, not the body, and "for-its-own-sake," attention is directed to the internal relations or "expressive properties" of the "work" rather than any "extra-musical" conditions external to it or that support its performance or

> use.... Thus, the connection of music with praxis (i.e., in terms of music's social purposes and values)—for example, with music as worship, celebration, ceremony, and so forth—either disqualifies it as properly aesthetic or considerably lowers its value on the tacit aesthetic hierarchy, depending on the theorist. (41)

The sonic-structural simplicity of spiritual expressions—such as a *Kyrie* or the Three Jewels—*lowers its value* in aesthetic scrutiny. But musics can be simple from an aesthetic point of view but have great spiritual value. Scholars theorizing spirituality in music education who might not identify as praxialists, like Boyce-Tillman, help realize and modify the praxial theory by identifying how it "connects the body/mind and spirit in musical experience" (Silverman et al. 2014: 62). A praxial approach to spirituality in music education on soil is employed, and aesthetic concerns can be understood as *within* but cannot *subsume* the broader spiritual praxis. In other words, while *aesthetic rationales are simply not needed*, an individual's aesthetic experiences can be a part of a spiritual praxis for many musickers. But if spirituality is about connecting (to God(ess)(s) however conceived, others, and nature), the individual's aesthetic experience is only a part of spirituality. What matters is what musickers intend, what they enact, and what they experience. This praxis is essentially social and plural, just as the self connects, extends, and includes.

Sentence 2: Musickers music for spiritual aims.

Toward an Ecotheology

Though, in the U.S. at least, Christianity and environmentalism seem at odds, Christianity is filled with ecological teachings. Critical of modern society, Christians recognize, "Our times are agitated and violent and threaten the future of humankind and our shared home, the Earth" (Boff 2001: 19). *Conversion*, central to Christian thought, is more than saying "I believe." Conversion is lived daily as critical conversation with society and the world. Conversion and conversation share Latin roots (*con* = with/together; *verse* = turning). "Christians are those who unceasingly ask critical questions of the society in which we live, and who continuously stress the necessity for conversion, not only of the individual but also of the world" (Jonas 2009: 177). The relationship between people and the environment is stewardship: "the responsible care of property belonging to another" (Berry 2010: 99). In belonging together, humans converse with each other and with non-human life; we change together—we convert.

"Look, the heavens, even the highest heavens, belong to the Lord, your God, as well as the earth and everything on it" (Deut. 10:14, NABRE). For Berry (2010), stewardship challenges our current economic system: "Large accumulations of land were, and are, forbidden because the dispossession and privation of some cannot be an acceptable or normal result of the economic activity of others, for that destroys a people as a people; it destroys the community" (99). Christians are called to convert; to become ecologically conscious stewards aware of society's unjust domination of people and our environment.

Correspondingly, Christianity and anarchism seem at odds. However, according to Christian anarchists, "biblical thought leads straight to anarchism—anarchism is the only 'anti-political political position' in harmony with Christian thought" (Ellul 2006: 157). Anarchists have been anti-religion to the extent that God has been conceptualized as "power" (God as master; humans as slaves), and Christianity as seeking order above freedom. Constructing a biblical argument, Jacques Ellul challenges institutionalized Christianity as antithetical to the God of the Bible, where Jesus "chooses the way of radical nonviolence" and "Love is the only 'definition' of God given in the Bible" (160). This position of nonviolent action and skepticism toward institutions provides agents (doers) with a biblically rooted foundation for acting and judging action.

Ellul (2006) continues, "As the institutionalization of the Church hardens, so does its dogmatism: truth considered as a possession[12] (in which case it ceases to be truth) leads to judgment and condemnation. Love when institutionalized produces authority and hierarchy" (160). Christian anarchists identify the institutionalization of values[13] (such as love) historically. For Illich (1971) schools exemplify the negative consequences of the institutionalization of love in industrial society. "School serves as an effective creator and sustainer of social myth . . . academic priests mediate between the faithful and the gods of privilege and power, a ritual of expiation which sacrifices its dropouts, branding them as scapegoats of underdevelopment" (44). When I act as a human, I tend to act in a morally good way. But when I act as a representative of an institution, a teacher, I too often feel the responsibility of reifying school's primary value—scarcity. As a teacher I construct objectives, order how students *should* learn, and I grade students—A, B, C, D, or F. If I give all of my students an "A" I feel I have failed my institution in some way. Not everybody can be an "A" music, or math, or reading student. In societies without school music, no students are "A" music students; but they also aren't "F" music students—they may not even be *students*, but simple, beautiful, learning/acting *musickers*.

Despite knowing the damage it does to people and to musical cultures, I have labeled students non-musicians, and they have become dropouts, dropping out of school music. I have had university students identify themselves as non-musicians, even students who perform as musicians without training. Cultures without music education institutions do not reify musician scarcity to the same degree, if at all. In human history, the vast majority of people have been musicians without music education. But the argument for music in schools becomes one of equal access of a modern culture that doesn't view music as important. But even in this modern culture, these non-musicians devote quite a bit of effort to experiencing musics.

Fromm (1969) recommends spontaneity[14] to counteract institutionalization of scarcity in schools: "*positive freedom consists in the spontaneous activity of the total, integrated personality*" (257, emphasis in original). Music educators can begin with skepticism toward their institutions, and incorporate activities that nurture student spontaneity because these creative activities empower the students as individuals and community members, rather than as subjects of institutions (e.g., schools, government/industry testing, the National Association for Music Education, the MayDay Group).

Fromm (1969) also suggests spontaneous activity aids in self-cultivation, connecting people to themselves and nature (259). For music teachers and music teacher educators, we can say that university music teacher educators do not prioritize these activities (Elpus 2013; Hewitt and Koner 2013; Stringham et al. 2016); music teachers can experience a tension between creative pedagogy and long-held repertoire-based concepts of success (Abramo 2009); that, with more instruction, music teachers would include creative pedagogies (Bernhard 2012; Stringham et al. 2016); but that lack of creative musicianship could be a barrier (Kratus 2007), as could gender (Wehr-Flowers 2006). Fromm's insight extends what we know about creative pedagogy to the realm of self-cultivation.

Incorporating opportunities for spontaneous pedagogy can be central to teaching for self-cultivation and connection to nature, "for it is in the spontaneous realization of the self man united himself anew with the world—with man, nature, and himself" (Fromm 1969: 259). This is in line with my historical research into the spiritual philosophy of Satis Coleman, which integrated creativity, singing, world music pedagogy, and spiritual aims such as living simply and wholesome humility (Shevock 2015), and with Hanh's (2013) Buddhist spirituality. These were integrated aspects of a holistic spiritual philosophy for music education.

Sentence 3: Musickers creatively music for spiritual aims.

With a Christian anarchist concept of what we can term spirituality portrayed, and some implications for music education clarified, I pivot to the tradition of Catholic ecotheology. St. Francis of Assisi, St. Hildegard von Bingen, Teilhard de Chardin, and Thomas Merton contributed to ecological dialogue, though Catholics are "not prone to think of these and other significant contributors from Catholicism as being particularly ecological"

(Carroll 2001: 169). As Pope Francis's encyclical, quoted at the opening this chapter, illustrates, Catholic ecotheology connects social and ecological concerns. This includes "the physical and the spiritual as well as the human dimensions of reality" (Berry 2009: 2).[15]

Catholic ecotheology can be understood as being rooted in the Bible. Jesus meditated in the desert wilderness before beginning his mission: "Jesus was led by the Spirit to the desert" (Mt 4:1 NABRE). And poverty, a virtue emphasized in Catholic teaching, is a virtue out of line with our current society. According to St. Hildegard von Bingen, "No one is able to be satisfied by abundance—you are only bored by it" (in Fox 2012: 101). Pope Benedict XVI (2010)[16] taught safeguarding the environment is imperative: "Protecting the natural environment in order to build a world of peace is thus a duty incumbent upon each and all" (6).

White (1967) identified St. Francis of Assisi as "The greatest spiritual revolutionary in Western history" (1207),[17] and it is probably no coincidence that the current pope has chosen Francis as his namesake. With the publication of *Laudato Si'* in 2015, Catholic ecotheology was further validated. Pope Francis (2015) draws on St. Francis's "Canticle of the Sun." He discusses pollution, climate change, water security, loss of biodiversity, and societal issues such as quality of life, inequality, and weak responses. Recommending an integral approach that considers ecological and social issues simultaneously, he identified education as central to solving the ecological crises: "There is a nobility in the duty to care for creation through little daily actions, and it is wonderful how education can bring about real changes in lifestyle" (138). Within this educational theology, music might have a central role:

> If someone has not learned to stop and admire something beautiful, we should not be surprised if he or she treats everything as an object to be used and abused without scruple. If we want to bring about deep change, we need to realize that certain mindsets really do influence our behavior. (140)

Drawing on Teilhard de Chardin, Thomas Berry began writing about ecological issues in the 1970s, and he's connected to the deep ecology movement (see Chapter 4, "Deep Ecology"). He founded the Thomas Berry Foundation with the aim of "enhancing the flourishing of the Earth community."[18] Berry (1999) theorized the Great Work, creating one of "those overarching movements that give shape and meaning to life by relating the human venture to the larger destinies of the universe" (1). Berry's work provides a broad spiritual praxis for understanding the cosmos as evolving creation and acting for change.

> *Sentence 4: Musickers creatively music for spiritual aims out of a sense of duty toward people and creation.*

Interconnecting Wisdom from Buddhist Ecotheology

I wasn't the first Catholic to find wisdom in Buddhist teachings; Thomas Merton and Thomas Berry both wrote books decades ago. This different—different from my Catholic upbringing—cosmology and practice provided insight into the nature of humans, nature, and God. Buddhist ecotheologians stress the importance of hope and active meditation exercises to realize a more sustainable, interconnected relationship with the environment. "We can choose life. Dire predictions notwithstanding, we can still act to ensure a livable world" (Macy and Brown 1998: 16).

Music educators can choose life in our work to recover the planet, acting in the small ways our profession provides. "Regardless of what work you do, part of our work is to help bring about a collective healing, transformation, and awakening for our own well-being and for the sake of our planet. The insight of interbeing can help in this, but we need a collective awakening" (Hanh 2012: 104). Accordingly, people and nature are interconnected (*interbeing*) in ways we may not be fully conscious of. "The radiant and elegant cosmos that we can observe is in fact our own consciousness itself and not something outside of it" (Hanh 2013: 12). Mindfulness, which in turn increases consciousness of interconnectedness and love for the earth, is practiced daily through meditating during usual, everyday activities (e.g., 86–96).

For deep ecologist and Buddhist Joanna Macy, despair caused by ecological crises requires a calming, spiritual response. When we begin to meditate we become aware our pain and nature's are one; "feelings of pain for the world arise" (Macy and Brown 1998: 114). They recommend a series of meditation exercises to gain insight into our wisdom, and power, to communicate, plan, and act for ecological change (see 169–181).

> *Sentence 5: Musickers creatively music for spiritual aims out of a sense of duty toward people and creation, in recognition of interconnection with other people and creation.*

Creative Vignette

My ear, defenseless and always open, rings. Stressed by the clang of society, by television pundits, internet trolls, and incessant news, by politicians bought and sold by big, polluting industry, and by family and friends corrupted by globalized denier discourse, my head spins. Weighed down by dissonance, I sink and gasp for breath. I'm unable to hear my voice in the cacophony. Urged by a favorite poet's words, "I go to the woods alone" (Oliver 2010: 5), I sit on a fallen maple, allow the

(continued)

(continued)

cool earth to penetrate my flesh, and breathe slowly. Unruffled by the harsh music, chipmunks play on leaves and dirt. On creation. I listen. I permit the discord to dissolve, allowing healing melodies of nature to hold my consciousness. In the soothing soundscape, I embrace the tree below and become invisible to myself in the great work. Recovery. A song emerges from the roots of the tree of life to the uppermost branches, and resonates throughout God's cosmos.

Viable Humans and Viable Music Education Praxis

Perhaps the greatest challenge facing people living in modern, industrialized society is viability. Spirituality begins by slowing down, reflecting, and recognizing what we can and cannot accomplish. Humans play a role in the cosmos (Berry 1999). But it is not necessarily Promethean (Illich 1971). Living humbly, music educators can help students realize a sustainable lifestyle by counteracting society's competitive elements (Shevock 2015: 59). To remain viable, the individual person must realize interconnections with other people, interconnections with nature, and develop a deeply historical perspective—what Thomas Berry (1999) called "the Earth story" (21). In this way, self-cultivation can become realization of a broader self (Naess 2006).

Arne Naess discusses self-realization, what I call "self-cultivation" (see Gossett and Shevock 2015). In line with Catholic ecotheology, Naess (2006) recognized the connection between economics and self-realization: "severe threats to economic interests correspond to possibilities of severely reduced self-realization" (93). He also recognizes the interconnected nature of people: "the absolutely highest level of self-realization cannot be reached by anybody without all others also reaching that level" (94). He also connects the recognition of others' *rights* to self-realization. Those rights are extended to non-human animals (bears, sheep, and wolves). He calls this *species egalitarianism*, "the equal right in principle of all species to live and flourish" (98). In music education, we can recognize the rights of non-human beings to music for self-cultivation within their own conception of it.

> *Sentence 6: Musickers creatively music for spiritual aims out of a sense of duty toward people and creation, in recognition of interconnection with other people and creation, and in acknowledgment of all beings' rights to self-cultivation.*

A Deep Ecology Musicking Spirituality

Deep ecology has a uniquely spiritual aspect. Musicking can be vital for spiritual praxis. For Devall (1988), writing music and poetry can cultivate an appreciation for our relationships to and the value of all living beings (73). Referring to Paul Winter, George Winston, Buffy St. Marie, and others, Devall suggests, "Music coming from the intuition of deep ecology paradoxically expresses the 'great silence.' Sounds resonate with earth and sky" (79). Going to nature is often described as finding silence. I explore this idea when I compose and perform ecological jazz, inspired by Paul Winter's compositions with whales, wolves, and other non-human animals. In this practice, I record non-human musicking; I modify it and use it to inspire composing beats, chord progressions, and melodies. Ecological jazz recognizes the intrinsic value of non-human musicking, the importance of place, and is a creative practice. As such, it is a viable spiritual expression of musicking on soil.

Paradox is a vital part of spirituality (Palmer 2008; Shevock 2015). Silence to sound is one paradox. To practice interconnected spirituality in our daily lives, we recognize disconnected and alienating aspects of global institutions and, as a result, many people living today. That which is outside of us is inside. Global change must begin by listening to our local places. This localization of change recognizes the importance of grassroots activity—people in community choosing not to live unsustainable, generalizable lives (Esteva and Prakash 1998). The primary modern paradox is that many people live unsustainable lives as consumerist cogs in a globalized industrial economy. The paradox is resolved with the realization that all generalized answers are oppressive to some people somewhere. In recognizing the oppression of global institutions, even ones acting for a *good*, people can react in ways that further damage their local places. For instance, police officers, particularly those who selected this profession to protect their local community, can find themselves defending global industries against their own community members who want less polluted water, or to keep a cherished mountain from mountaintop removal. Only by empowering local people to listen to their soil, global ecological crises like climate change can be resolved. This listening is ecological (interconnected/interdependent) and spiritual/religious (like ligaments of a body).

Because of our work with the sounds and silence of creation and creating communities, music educators reside in a distinctive position to begin this work. This involves acting with skepticism toward institutions we work for. For music educators with one foot in the schoolroom and the other in the community, empowering people and challenging institutions is possible. I would think businesspeople don't lop off mountaintops in their hometowns for marginally greater profit in coal extraction. Schools shouldn't lop off local musics. Generalized values, like those held by shareholders and education policy makers, are needed to commit such ghastly wrongdoings. If music education will be viable, if we are going to cultivate viable humans, it should be spiritually reconnecting, localizing, and skeptical of institutions, even ones that aim in theory to do good.

> *Sentence 7: Holding the potential paradox of internal and external silence and sound, of working for and against our institutions, musickers creatively music for spiritual aims out of a sense of duty toward people and creation, in recognition of interconnection with other people and creation, and in self-cultivation and acknowledgment of all beings' rights to self-cultivation.*

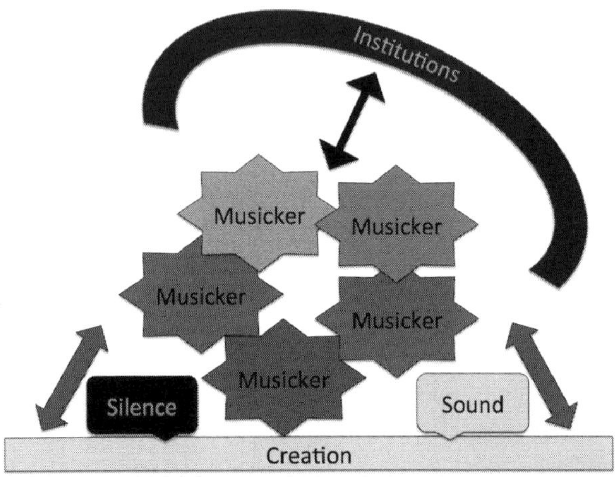

Figure 5.1 A Spiritual Praxis

Sounding Wild, Hearing Wild

all that thrives within the mind
  all that loves, lives, and binds
  all that moves; all that waits
  all that's cruel and kind
comes from old and ancient oak
  comes when Gabriel's horn first spoke
  comes asleep when dreamers woke
  wakening to truth non-remote
windy, chilling wilderness
  charming, sunny wilderness
  thunder crashing, wilderness
  dying stags of wilderness
  living wild
  loving wild
  sounding wild
  hearing wild

Notes

1 From the 2015 encyclical *Laudato si'* (35, emphasis in original). Pope Francis seems critically important for the redemption of 21st-century Christianity. As Boff (2014) writes,

> Suddenly, as if by magic, Pope Francis appears on the scene of history. He comes from 'the ends of the Earth,' the Third World, Argentina, outside the tensions of the Roman and European church, and he inaugurates another style of being pope and being church. There are no better words to describe this than the terms *break* and *beginning anew (inauguración de lo nuevo viable)* (7).

2 Link: http://www.pbl.nl/sites/default/files/cms/publicaties/500114012.pdf.

3 This sentence—*Musicker musics for spiritual aims*—is constructed to parallel the phenomenological form: ego (a musicker) cogito (musics) cogitatum (spiritual aims) (see Husserl 1983). This form is extended as sentence structure throughout this chapter: to 2) *Musickers music for spiritual aims*; 3) *Musickers creatively music for spiritual aims*; 4) *Musickers creatively music for spiritual aims out of a sense of duty toward people and creation*; 5) *Musickers creatively music for spiritual aims out of a sense of duty toward people and creation, in recognition of interconnection with other people and creation*; 6) *Musickers creatively music for spiritual aims out of a sense of duty toward people and creation, in recognition of interconnection with other people and creation, and in acknowledgment of all beings' rights to self-cultivation*; and 7) *Holding the potential paradox of internal and external silence and sound, of working for and against our institutions, musickers creatively music for spiritual aims out of a sense of duty toward people and creation, in recognition of interconnection with other people and creation, and in self-cultivation and acknowledgment of all beings' rights to self-cultivation.*

4 Fromm (1976) connects his having/being modes to teaching and learning. "Optimum knowledge in the being mode is *to know more deeply*. In the having mode it is *to have more knowledge*" (41). If music teachers and learners are going to address the serious crises we face, we need knowledge, but we (collectively but not many individuals) already know enough to act. It is time for us to change our *being*, so as to live sustainably on the planet.

5 Catholic liberation theologian Leonardo Boff (2001) identified religion as central to connection: "Experiencing this feeling of belonging to a greater whole means feeling umbilically connected and re-connected [*re-ligado*] to the divine heart that fills life with peace and serenity" (33). According to Boff, the word "religion" refers to a body, a reconnection of ligaments.

6 Parker Palmer (2008) writes,

> Only when we lose our connection with one another do we need a code of conduct to tell us what we ought to do. When life becomes fragmented, our organic responsiveness to one another is replaced by 'oughts.' Eventually these oughts become a system of abstract thought far removed from human need, a creed to be defended rather than a relationship to be lived (24).

7 Coleman's use of language (such as "primitive man") reflects the invisible gender bias and unjust power relationships common to early 20th-century writing. As a

music educator, Coleman seemed to make some important connections between music, nature, and deity(ies) that are not present in much contemporary music education scholarship.

8 Van der Merwe and Habron's (2015) conceptual model for spirituality in music education employs van Manen's lifeworld existentials, "corporeality, relationality, spatiality, and temporality" (47), promoting a phenomenological understanding of spirituality that can be congruent in many ways with the spiritual praxis proposed. In line with models of other researchers in music education, their model distinguishes between *spirituality* and *religion*, something the current spiritual praxis does not attempt. In the current chapter, religion is defined differently than in this previous research. Extending the ideas of Catholic writers like Leonardo Boff and Henri Nouwen, because of the connections religion makes among people, the dichotomy between spirituality and religion seems unnecessary. I suspect distinguishing the terms (spirituality and religion) is meant to help people who have had negative experiences with specific religions (communities that might be too socially conservative for some people). I can appreciate those concerns. Religious institutions face many of the same corrupting influences as educational institutions, and should be challenged in good faith.

9 Matsunobu's (2011) study of North American students of Japanese shakuhachi music found students were more interested in this universal human experience than "culture-specific dimensions of music" (284). This seems to place spirituality at odds with praxialism. However, North American students' desire for universal experience, as opposed to coming to understand culture, does not have to be understood as a *good*. Western society has long had a negative impact on non-Western cultures, and universalizing has been a major negative habit of Western peoples upon the social majorities (Esteva and Prakash 1998). Understanding the importance of paradox (Palmer 2008) to spirituality, it may be possible for music educators to hold these desires for universal experiences (traditionally the purview of aesthetics theorists) in check with a deeper respect for the importance of cultural difference in religious and spiritual experience. As Henri Nouwen writes,

> In the solitude of the heart we can truly listen to the pains of the world because there we can recognize them not as strange or unfamiliar pains, but as pains that are our own. There we can see that what is most universal is most personal and that, indeed, nothing human is strange to us.
>
> (Jonas 2009: 176).

10 Boyce-Tillman (2007) begins her chapter with five questions; these two seem the most pertinent to cultivating a spiritual praxis of music education on soil. It is interesting also that Boyce-Tillman conducts interviews around a piece, *The Healing of the Earth*, which has distinctively ecological aims.

11 In this case "not-religion" can be loosely constructed as spiritual activity intended by a non-theist (e.g., an atheist, secular humanist, or agnostic). Not all non-theists enact spirituality, but some do. For the construction of this spiritual praxis, this is categorized as an intention toward "religion." However, I'm not suggesting atheism, secular humanism, or agnosticism is a religion.

12 Here Christian anarchism connects to Fromm's (1976) psychology, where the *having* mode is understood as unhealthy in comparison to the *being* mode of existence.

13 Thus the Church was the joyous outcome of the unity of believers confident of their salvation, as they met together and showed forth God's love. But it became a structure, a custodian of authority and truth, representing God's power on earth. 'No salvation outside the Church' originally meant that those who recognized that Jesus Christ had saved them met together to give thanks (thus, outside the Church, there were no people living this faith). But the phrase came to mean that all those who are outside the framework of the Church are doomed! This reversal of meaning is quite a serious matter.

(Ellul 2006: 160–161)

In Chapter 4, "Deep Ecology," I shared a bit of Illich's analysis of sermons on "The Good Samaritan," which comes to the same conclusion about the institutionalization of Christian values having negative consequences.

14 St. Hildegard von Bingen's idea of "greening power" also connects creativity to ecological thought (Fox 2012: xv).

15 Or as former Environmental Protection Agency official Deeohn Ferris puts it, "In the metaphor of a rapidly sinking ship, we're all in the same boat, and people of color are closest to the hole" (in Lavelle and Coyle 1992: 3).

16 It was Pope Benedict XVI who recognized St. Hildegard von Bingen as a doctor of the church in 2012, validating her theological teachings. While the current Pope Francis is generally understood as representing more progressive elements, Benedict XVI is considered a conservative. In Catholicism, ecological sustainability isn't a liberal or a progressive issue, but a challenge for all.

17 White (1967) continues,

> Saint Francis, proposed what he thought was an alternative Christian view of nature and man's relation to it: he tried to substitute the idea of equality of all creatures, including man, for the idea of man's limitless rule of creation. He failed. Both our present science and our present technology are so tinctured with orthodox Christian arrogance toward nature that no solution for our ecologic crisis can be expected from them alone. Since the roots of our trouble are so largely religious, the remedy must also be essentially religious, whether we call it that or not (1207).

18 Link: http://thomasberry.org/.

References

Abramo, Melissa Natalie. 2009. "The Construction of Instrumental Music Teacher Identity." PhD dissertation, Teachers College, Columbia University.

Benedict, Pope, XVI. 2010. "If You Want To Cultivate Peace, Protect Creation." Message of His Holiness Pope Benedict XVI for the Celebration of the World Day of Peace. Online: https://w2.vatican.va/content/benedict-xvi/en/messages/peace/documents/hf_ben-xvi_mes_20091208_xliii-world-day-peace.pdf.

Bernhard, H. Christian, II. 2012. "Music Education Majors' Confidence in Teaching Improvisation." *Journal of Music Teacher Education* 22, no. 2: 65–72. DOI: 10.1177/1057083712458593.

Berry, Thomas. 1999. *The Great Work: Our Way into the Future*. New York: Bell Tower.

——. 2009. "Spiritual Tradition and the Human Community." [Originally published in 1987.] In *The Christian Future and the Fate of Earth*, edited by Mary Evelyn Tucker and John Grim, 1–7. Maryknoll, NY: Orbis.

Berry, Wendell. 2010. *What Are People For? Essays*. Berkeley, CA: Counterpoint.

Boff, Leonardo. 2001. *The Prayer of Saint Francis: A Message of Peace for the World Today*. New York: Orbis Books.

——. 2014. *Francis of Rome and Francis of Assisi: A New Springtime for the Church*. New York: Orbis Books.

Boyce-Tillman, June. 2007. "Spirituality in the Musical Experience." In *International Handbook of Research in Arts Education*, Part 1, edited by Liora Bresler, 1405–1421. Dordrecht: Springer.

Carroll, John E. 2001. "Catholicism and Deep Ecology." In *Deep Ecology and World Religions*, edited by David Landis Barnhill and Roger S. Gottlieb, 169–191. Albany: State University of New York Press.

Coleman, Satis N. 1928. *Bells: Their History, Legends, Making, and Uses*. Chicago, IL: Rand McNally & Company.

Devall, Bill. 1988. *Simple in Means, Rich in Ends: Practicing Deep Ecology*. Layton, UT: Gibbs Smith.

Ellul, Jacques. 2006. *Jesus and Marx: From Gospel to Ideology*. Eugene, OR: Wipf & Stock.

Elpus, Kenneth. 2013. "Music in the U.S. Federal Education Policy: Estimating the Effect of 'Core Status' for Music." *Arts Education Policy Review* 114, no. 1: 13–24: DOI: 1080/10632913.2013.744242.

Emmott, Stephen. 2013. *Ten Billion*. New York: Vintage Books.

Esteva, Gustavo, and Madhu Suri Prakash. 1998. *Grassroots Post-Modernism: Remaking the Soil of Cultures*. London: Zed Books.

Fox, Matthew. 2012. *Hildegard of Bingen: A Saint for Our Times*. Vancouver: Namaste.

Francis, Pope. 2015. *Laudato Si': On Care for Our Common Home*. Huntington, IN: Our Sunday Visitor Publishing.

Fromm, Erich. 1969. *Escape from Freedom*. New York: Henry Holt and Company.

——. 1976. *To Have or to Be?* New York: Harper & Row.

Gossett, Jason B., and Daniel J. Shevock. 2015. "Growing Inclusive Music Education Preservice Teachers through Self-Cultivation." Paper presented at the Society for Music Teacher Education Symposium, Greensboro, NC.

Hanh, Thich Nhat. 2012. *Work: How to Find Joy and Meaning in Each Hour of the Day*. Berkeley, CA: Parallax Press.

——. 2013. *Love Letter to the Earth*. Berkeley, CA: Parallax Press.

Hewitt, Michael P., and Karen Koner. 2013. "A Comparison of Instrumental Music Methods Course Content at NASM-Accredited Institutions." *Bulletin of the Council for Research in Music Education* 197: 45–61.

hooks, bell. 2003. *Teaching Community: A Pedagogy of Hope*. New York: Routledge.

Husserl, Edmund. 1983. *Ideas Pertaining to a Pure Phenomenology and to a Phenomenological Philosophy*, First Book, translated by F. Kersten. Dordrecht: Kluwer Academic Publishers.

Illich, Ivan. 1971. *Deschooling Society*. London: Marion Boyars.

Jonas, Robert A., ed. 2009. *The Essential Henri Nouwen*. Boston, MA: Shambhala Publications.

Jorgensen, Estelle R. 2003. *Transforming Music Education*. Bloomington: Indiana University Press.

Knieter, Gerard L. 1971. "The Nature of Aesthetic Education." In *Toward an Aesthetic Education*, edited by Bennett Reimer, 3–19. Washington, DC: Music Educators National Conference.

Kratus, John. 2007. "Music Education at the Tipping Point." *Music Educators Journal* 94, no. 4: 28–32. DOI: 10.1177/002743210809400403.

Lavelle, Marianne, and Marcia Coyle. 1992. "Unequal Protection: The Racial Divide in Environmental Law: A Special Investigation." *National Law Journal* 15, no. 3: 1–7. Online: http://www.ejnet.org/ej/nlj.pdf.

Macy, Joanna, and Molly Young Brown. 1998. *Coming Back to Life: Practices to Reconnect Our Lives, Our World*. Gabriola Island: New Society.

Martyn, Patricia, and Eric Brymer. 2016. "The Relationship between Nature Relatedness and Anxiety." *Journal of Health Psychology* 21, no. 7: 1436–1445. DOI: 10.1177/1359105314555169.

Matsunobu, Koji. 2011. "Spirituality as a Universal Experience of Music: A Case Study of North Americans' Approaches to Japanese Music." *Journal of Research in Music Education* 59, no. 3 (October): 273–289.

Naess, Arne. 2006. "Self-Realization in Mixed Communities of Humans, Bears, Sheep, and Wolves." *Trumpeter* 22, no. 1: 89–99. [Originally published in 1979.]

Nasr, Seyyed Hossein. 1990. *Man and Nature: The Spiritual Crisis in Modern Man*. London: George Allen & Unwin. [Originally published in 1968.]

Oliver, Mary. 2010. *Swan: Poems and Prose Poems*. Boston, MA: Beacon Press.

Palmer, Parker J. 2008. *The Promise of Paradox: A Celebration of Contradictions in the Christian Life*. San Francisco, CA: Jossey-Bass.

Regelski, Thomas A. 2016. *A Brief Introduction to a Philosophy of Music and Music Education as Social Praxis*. New York: Routledge.

Shevock, Daniel J. 2015. "Satis Coleman—A Spiritual Philosophy for Music Education." *Music Educators Journal* 102, no. 1 (September): 56–61. DOI: 10.117/0027432115590182.

Silverman, Marissa, Susan A. Davis, and David J. Elliott. 2014. "Praxial Music Education: A Critical Analysis of Critical Commentaries." *International Journal of Music Education* 32, no. 1: 53–69. DOI: 10.1177/0255761413488709.

Stringham, David A., Linda C. Thornton, and Daniel J. Shevock. 2016. "Composition and Improvisation in Instrumental Methods Courses: Instrumental Music Teacher Educators' Perspectives." *Bulletin of the Council for Research in Music Education* 205: 7–25.

van der Merwe, Liesl, and John Habron. 2015. "A Conceptual Model of Spirituality in Music Education." *Journal of Research in Music Education* 63, no. 1: 47–69. DOI: 10.1177/0022429415575314.

Wehr-Flowers, Erin. 2006. "Differences between Male and Female Students' Confidence, Anxiety, and Attitude toward Learning Jazz Improvisation." *Journal of Research in Music Education* 54, no. 4: 337–349.

Weil, Simone. 1952. *The Need for Roots*. New York: Routledge.

White, Lynn, Jr. 1967. "The Historical Roots of Our Ecologic Crisis." *Science* 155, no. 3767 (March): 1203–1207.

Postlude

> You don't get to choose how you're going to die. Or when. You can only decide how you're going to live. Now.
>
> (Joan Baez, 1968[1])

Dirty Work

On November 8, 2016 Donald J. Trump was elected President of the United States. The U.S. is one of the largest forces of ecological destruction on the planet, as well as a hope for global change—this is the nation that invented the wilderness preserve. But the election doesn't bode well if the U.S. must lead people to live more sustainably. An open climate change denier, Trump suggested snowstorms prove global warming is a hoax (Boykoff 2013: 809). A month later, President-elect Trump chose climate change denier Scott Pruitt to head the Environmental Protection Agency, and Exxon Mobil CEO Rex Tillerson, perhaps not a denier but still a challenging pick for people concerned about the ecological crises, to be Secretary of State. Deniers misrepresent climate science as in doubt, and grab headlines. The "disproportionate coverage of outlier contrarian, skeptical, and denialist claims . . . has challenged efforts that seek to expand rather than constrict the spectrum of possibility for varied forms of climate action" (811). In recent years, ecological sustainability has become a so-called *liberal* concern, and not a concern for all humanity. More than a failure in media coverage, Trump's election may point to an underlying failure of education institutions to cultivate ecologically literate citizens and, deeper, a problem of the human spirit—a religious and spiritual problem, as suggested in Chapter 5, "Spiritual Praxis."

There was little discussion of climate change or any other ecological crisis during the 2016 U.S. presidential campaign. Climate change wasn't a significant portion of any of the debates; wasn't a driving headline for paper, online, or television media; and didn't seem a part of day-to-day conversations about the election in my local community. This was painful. Those who voted for Trump are often described as doing so for economic reasons—poor, White, rural voters who were left behind by the global neoliberal economy. This seems like a failure of consciousness of intersectionality—"the relationships among multiple dimensions and modalities of social relations and subject formations" (McCall 2005: 1771). Both major parties failed to realize intersectionality, discussing each issue separately, and severed from others. Intersectionality proposes gender, racial, economic, and ecological issues can be analyzed as related—that is, intersecting. If intersectionality is accurate, it is impossible to improve social concerns without addressing the ecological crises.[2] Intersectionality exists because there is an underlying reality, that of unjust domination (see Chapter 4, "Deep Ecology").

If intersectionality is false and economic gain is opposed to ecological sustainability, which is intimated in much popular rhetoric, then the challenge of human survival goes to the root of our neoliberal economy. This gives grave meaning to *dirty work*—jobs are inherently *dirty*, polluting the environment. However, given the choice of sustaining our unjust version of the free market economy or sustaining human existence the choice is obvious: human life is a higher requirement than our current economic system, whatever its positive elements. If humans cannot live, no economy exists. In this book, I have argued for an intersectional concept; environmental issues are economic, racial, cultural, and gender issues (see Chapter 4, "Deep Ecology"). And deeper, the ecological crises are a crisis of spirituality, a failure of religion (see Chapter 5, "Spiritual Praxis").

The intersections between the ecological crises, racial injustice, patriarchy, and economic stratification seem invisible to the educated public. For instance, in the Syrian refugee crisis, which began as unprecedented drought, we see an ecological crisis (climate change) intersecting with racial injustice, patriarchy, and economic stratification. Could an eco-literate citizenry reject the overwhelming evidence for human-caused global climate change, or care so little about the ecological crises to ignore these crises' cost to people (a cost felt most strongly by poor people and women), non-human animals, and plants?

This book is based on the premise that music educators have opportunities to educate students for ecological consciousness by orienting people to their local sonic environments and cultures, and cultivating global ecological literacy. Music provides unique ways to think about and act to address the ecological crises, and to teach and learn sustainable living. There are opportunities because of connections between the soundscape and musicking, and the people musicking for ecological activism, when music educators don't overlook these musics.

A Conservative

Driving to work, I pray the rosary. I learned the first three "Hail Marys" recall the three foremost virtues of Christian life: *faith*, *hope*, and *love* (*caritas* is also translated as *charity*). These virtues help me face the abyss of the ecological crises with courage.

In my childhood an old priest once said, "Faith isn't belief. If you believed, you wouldn't need faith." The nuances distinguishing the two concepts—*faith* and *belief*—often elude me. But I do often lose belief: in God, in myself, in my family, friends, society, and music education. Faith is embedded in the paradox of the cross—"the greatest paradox of all: that to live, we have to die" (Palmer 2008: 31). Faith reminds me in times of disbelief that I'm not God or even a god, and do not and cannot know many things. Seeing no reason to believe it's true, I have faith God will help humanity become better stewards of our common home. "On the cross, we give up the fantasy that we are in control, and the death of this fantasy is central to acceptance" (46). Perhaps faith begins with acceptance, especially in moments of incredulity, keeping us from submerging into psychological darkness. But it is also humility. Faith keeps my Promethean self, who would burn the whole world to steal from the gods, in check. Faith helps me to say *yes* sometimes, and at others to say *no*. This is sophisticated conscientization.

I have hope, not only that Jesus will return, but also that we can live better on this planet in the meantime. I hope we start to value all people, and not only rich White men. I hope we recognize how our society is unfair, and make it better. I hope we recognize our interdependence with non-human life. Hope is not immobile wanting. Fox (2012) discusses hope as a verb, recommending, "it's time to go to work to defend Mother Earth" (41). I hope as an action in society.

And finally love, more than other virtues, is a practice. The practice of love lies in the daily small acts making the world a better place.

> A community consists not of specialized professionals but of generalized amateurs. It is worth recalling that the root meaning of *amateur* is "lover," one who acts out of love for what one is doing. Love is ultimately the source of all abundance in life, and when love flows from us, our abundance becomes clear.
>
> (Palmer 2008: 112)

I undeniably identify as liberal. But expressing my philosophy within the context of prayer and Christian virtue makes me seem conservative. Palmer (2004) writes, "We project our inner darkness on others, making 'enemies' of them and making the world a more dangerous place" (16). I wonder how much accepting the preset labels *liberal* and *conservative* makes enemies of others.

Some of the values I hold dear can be understood as conservative—especially conserving soil, the places and traditions as lived for generations. A spiritual praxis is facing, not hiding from, paradox.

I am not conservative in the same way as Roger Scruton (2012), who theorizes an environmental conservatism. Though his version of conservatism could provide an expression of eco-literate music pedagogy, I recommend something different than more research, technology, and "decentralized initiative" (408) to solve the ecological crises. I recommend fundamental re-localization—reorientation to place, people, and local histories—unalike to his environmental conservatism. And I recommend a spiritual renewal—a philosophy of music education on soil. Identifying myself as conservative can be understood as a practice of love. It will help to clarify the ways I am connected to those I have least identified with. With an appreciation for diversity, it may be that *good* change lies in our commonalities. To identify myself as conservative requires reflection on what the word *conservative* inspires me to think about.

Conservatism begins in my role as an autoethnographer. As an autoethnographer, a scholar who analyzes and writes honestly about myself within the context of culture and history, my first aim is to conserve myself. That is, I want to conserve my humanity. That requires conserving many other things, including my flawed perspective, the humanity of others, the health of traditional cultures, and the rights of God's creation.

Conserving the humanity of others, health of traditional cultures, and rights of God's creation makes me conservative in a distinctive way, placing me in opposition to one of modern society's least conserving institutions, global neoliberalism. Though in the U.S. at least conservatives seem to be industry's greatest advocates, the term *neoliberalism* relates to *liberalism* for a reason. Neoliberalism extends liberal human rights to corporate persons. For instance, neoliberal ideals defend corporate persons' rights to move across national borders in a way many liberals wish human persons could. Neoliberalism doesn't discriminate openly. If a person of color, a woman, a transgender person, or a Muslim becomes a CEO, neoliberalism values them (paying them more and more while paying other workers less and less). Neoliberalism uses money to assign value; everybody in a neoliberal society knows how valuable s/he is—having an objective, nondiscriminatory measure of self-value.

Neoliberalism allows the wealthy consumers in America and Europe to buy products made in the two-thirds world and to claim lowered greenhouse gases, while declaiming the increased emissions of the two-thirds world. The lifestyle of neoliberal consumers is at the expense of the social majorities. Neoliberalism is anti-conservative in destroying the religious traditions and places of many cultures and rural people, declaring their children's *right* to be educated, to be developed, to be urbanized, or better yet suburbanized, since urban areas aren't nearly as polluting as rich suburbs (see Brown 2014). Neoliberalism teaches the next generation to leave home, and the young to think of themselves as more intelligent, more technologically savvy, in short *better* than their bumpkin parents. This is consistent with Bowers's (2007) comment, "only in America is free-market liberalism labeled as the expression of conservatism" (ix).

Neoliberalism's inequities aren't in its conservatism. Neoliberalism isn't about conserving anything really, and it's a very new economic system compared to other historic market systems. Neoliberalism's inequities are in its extension of liberalism. It's no surprise that our neoliberal music education institutions value suburban music education programs, and fill university positions with those who were "successful" (e.g., winning performance competitions, something music teachers in poor rural and urban places can hardly hope to accomplish with any regularity) in preset, antiquated, and rigid job categories—many urban and rural music educators teach bands, choirs, orchestras, general music classes, and others; they don't focus on one. This is understood as a lack of expertise. In many rural communities band prevails, but without the prestige (such as competition trophies) suburban band directors receive. The neoliberal suburbanites, selling our children to the gods of capital—the music industry, the technology industry, the record industry—define their work as success (see Illich 1968), even as their values destroy the earth, and then go into neoliberal universities and train the next generation of hopeful music teachers to be as neoliberal, as technologically savvy, as critical of poor social majorities, and as polluting.

I am conservative. To conserve is to serve (its Latin root, *servāre*); and like conversion and conversation, this occurs with (*con*) other human and non-human life. In conversation I convert to sense what to conserve. I want to conserve musical traditions that cannot and should not be ossified and petrified in universities. I want to conserve places and musics that industry doesn't. I want to leave a viable, diverse, healthy planet for my child. This may be conservative, but seems contrary to a conservative movement hijacked by neoliberal ideology. Universities and other industry have a deleterious history of conserving culture, including languages and musics (Prakash and Esteva 2008). Only *dirty work* (living on soil), within the reach of agents who know each other, can conserve places, traditions, and history.

Musicking on Soil

Conserving the soil, in faith, hope, and love I look for God's plan in church, in the Bible, and in creation. The history of farming practice in the 20th century can be typified as the history of industrial fertilizer use. Industry-produced fertilizer is made primarily of three nutrients soil needs—nitrogen, phosphorus, and potassium—increasing crop yields. Today, fertilizers add other secondary macronutrients and micronutrients. During the 20th century, gardeners recognized that industrial farming practices were negatively impacting their garden vegetables. Soil became a dead thing that only served to hold plants upright. These gardeners (and some industrial-scale farmers) revived the practice of composting. Farmers composted for centuries, but the promise of industrial progress led them away from the practice. Compost is an inefficient fertilizer. Compost includes the same three nutrients as fertilizer in smaller quantities, but also, because it is made from decaying matter, includes many others, as well as living organisms (e.g., earthworms, bacteria, protozoa, fungi). And compost helps form humus, aerating the soil (yes, roots breathe air).

Conserving soil, the lesson it teaches is the essential place of diversity in structural health. Healthy soil requires calcium, copper, boron, and carbon; actinobacteria and rotifers; water, air, and even squirrels. And healthy organizations require diversity (of gender, race, ethnicity, politics, religious perspective). Increased yields may result from using less diverse fertilizers, but these yields are temporary and can lead to soil depletion. Soil requires the informed work of people. People compost. People apply the compost to their gardens and farms. People work to make institutions more diverse, and weed them regularly of inequities. The conservative position can be understood as anti-institutional, a position that takes on more complicated significance since the election of Donald J. Trump. Perhaps music educators can learn that our institutions, even our best ones, need to be composted, tilled, watered, and weeded regularly. Sometimes our current institutions need to be discarded. The compost pile turns dead matter into new life.

Musicking on soil is *dirty work*, a place-conscious beginning. Teachers and students listen to the soundscape—where we live, now. People living unsustainably in their local places cannot model global aims, however noble. This is rootless, unjustifiable wishing. We cannot ask others to do what we are unwilling to do. Beginning with institutional skepticism, countless stores have asked consumers to sacrifice an extra dollar at the cash register and claim to have donated money without denting profit margin. The Nature Conservancy allows oil drilling on the Texas City Prairie Preserve as the critically endangered Attwater's prairie chicken disappears (Gillis 2014). All institutions, however noble, have earned our watchful skepticism. Music educators living on soil begin to hold themselves to higher standards, even while imperfectly accomplishing sustainable living and musicking.

Cultivating Eco-Literate Musickers

Eco-literate music pedagogy begins on soil, place-conscious, and soundscape focused. The music educator focuses on creative pedagogies that respect local cultures, histories, and ways of life. However, many of the ecological crises are global. A music teacher in Pennsylvania may never see the Great Pacific Garbage Patch, but is called to understand it, to brainstorm solutions with his/her students, or to write songs. Experiences with local crises, such as fracking, might help Pennsylvanians think about the Great Pacific Garbage Patch. While each crisis is unique in some ways, the underlying economic, social, cultural, historical, gendered, and spiritual roots can be identified.

Ecological literacy attempts to understand others, especially non-human others where people do not live. The musics of whales whose waters are polluted by deep-water drilling, or polar bears trying to survive Arctic ice melt can be an entryway for cultivating ecological literacy. The musics of non-Western cultures, whose music can embed alternative ways of thinking about people and the environment (see Seeger 2016), also cultivate ecological literacy.

Questions

There were four questions guiding this autoethnography. The first question was: *What does it mean to embrace the impulse to teach music* for *ecological literacy?* It seems inevitable that I would theorize eco-literate music pedagogy. The seeds for this were present in my earliest experiences, especially in reflecting on my grandfather working to build parks, and my dark night of the soul leading to a more experienced spirituality. However, the lack of ecological *ends* for music education prompted hesitation. Who am I to theorize something so radically out of step with modern music education praxis? The work of Satis Coleman and more recent music educators such as Vincent Bates helped me overcome hesitation.

The second question was: *What is it like to theorize eco-literate music pedagogy?* Because of incongruence with modern music education, I returned to my core spiritual beliefs—rooted in Catholic ecotheology, consistent with deep ecology, skeptical of institutions, place-conscious, and looking to soil (the "book of nature") for wisdom. The third question was: *What is learned through enacting this pedagogy?* I learned that there are a wide variety of creative pedagogical options afforded by the soundscape; that composers have been inspired by nature's music for centuries; that people have used music in ecological activism; and that these afford opportunities to consider the ecological crises as expressions of an unjust society.

The final guiding question was: *How do the impulsion, the theorizing, and the enacting relate to one another?* Through experiencing the impulsion, the theorizing, and the enacting of eco-literate music pedagogy, I gained insight into my values as a teacher; how music teaching and learning relate to society positively and negatively; what types of practices can help society meliorate; and the importance of rooted, specific praxis, as well as skepticism toward universalizables.

Will Music Education Change?

Music education scholars, with some exceptions (Bates 2013; Koza 2006; Morton 2012; Shevock 2015), have not expressed a music education theory congruent with sustainability or for ecological literacy. Perhaps this is our moment, the early 21st century. There is cause for hope, a reason this book is being published now. Berry (2009) suggested humankind is experiencing a "new ecological consciousness" facilitating "a global scale of spiritual awakening" (6). Eco-literate music pedagogy provides a model for sustainable teaching praxis, as we work against the forces of "great darkness" (6) currently devastating our common home. Music educators at the forefront of this work are charged to increase ecological consciousness: consciousness rooted in local places, local histories and ways of knowing, and local challenges to a sustainable environment. They are also challenged to see global crises that may not be experienced locally. The current book represents one possible praxis for music education as ecologically literate. It characterizes a philosophy rooted in soil, locally focused, and acknowledges the advantages of diversity.

If this philosophy of music education on soil becomes a stimulus for increased eco-literate teaching in school and community music—e.g., general music classes, choir, band, orchestra, and general classes that employ music—it will be successful. If eco-literate music pedagogy inspires other philosophies of music education that provide a clearer path to attending to our ecological crises, it will have accomplished its task. If however, in five or ten years, music educators and music education scholars continue to ignore the ecological crises, this book will have failed. Can music education change? I believe it must. Will it? That question must suspend unanswered for the time being. I am hoping. I am listening for resolution.

New Directions for Research

Various research methods can deepen our understanding of eco-literate music pedagogy and a philosophy of music education on soil. Surveys can show how much and what types of eco-literate activities music teachers are currently using. Descriptive case studies can illustrate how music classrooms are incorporating soundscape pedagogy, or writing music for ecological activism, and can deepen our understanding of pedagogy. Interview studies can spur speculation about ecological issues and invoke music practices that had not previously been considered. Experimental researchers can identify various music teaching and learning techniques and test which ones are more effective at cultivating ecological literacy. Because this book characterizes a new philosophy of music education, much research can be done. The more researchers consider these issues, the more likely a viable eco-literate music pedagogy will be constructed, modified, and enacted.

What Does a Philosophy of Music Education on Soil Help Us to See?

To begin with, the current philosophy of music education on soil helps music educators see and address a number of social movements centering on the invitation to live and music sustainability, such as indigenous ecological justice, and slowing climate change, waste, and species collapse. This begins with an inclusive definition of the verb *music* as *the intentional experiencing of sound* (Chapter 2, "Ecological Literacy"), recognizing the intrinsic value of human and non-human being and musicking (Chapter 4, "Deep Ecology"). A philosophy of music education on soil is not *only* a tool for facing the ecological crises, but also a deep-seated conversation/conversion/conservation. By beginning where our profession has been deaf, music education on soil opens us to hear many under-critiqued concerns. These include, but are not limited to, the techno-utopian view (the unsophisticated perspective that new devices are automatically and always good), how universalizables in methods books, repertoire, and curricula can negatively affect local ways of knowing, the oppressive nature of institutions such as schools, the music industry, professional organizations, and counter-movements (there are no "sacred cows," because institutions can only improve by being challenged), the value of non-institutionalized human relationships, the value of religious and spiritual traditions, and the value of family musicking, often ignored in school music.

Music education on soil respects the virtues of soil, such as diversity, longevity, place, patience, nature, intergenerational living, and love. Music education on soil begins by humbly listening to place—to human and non-human beings that have lived sustainably in place for years, decades, centuries, and millennia. Music education on soil emerges as a humble philosophy of specifics, and not generalizations. In farming, cultivating, and improving, clay soil differs from sandy soil; plains from mountain; inland from shore; city from suburb from rural. A gardener is as likely to take advice from a gardener down the street as from a farming scientist. By taking on the perspective of music education on soil, scholars can move the profession in more ethically *good* directions by looking downward, below our feet, where virtue is plowed, composted, planted, watered, and, with respect for the Epimethean work of sunshine and time, cultivated, birthed, and grown (see Illich 1990).

People living on soil value long-term, slow, sustainable solutions to today's fast-paced, uprooted, unsustainable challenges. On soil, we are skeptical of techno-fixes to the problems that our technological, industrial, consumerist society generated. In many ways, this is out of step with the quick, fast-paced, ecologically destructive answers society gives us. For instance, a short-term, fast-paced, ecologically destructive fix for the climate crisis might involve stratospheric sulfate aerosols. Once done once, they would need to be reapplied regularly ("rapid warming when it stops"). They are challenged by many climate scientists as risky, leading to possible ozone depletion, continued ocean acidification, enhanced acid rain, counteracting the efficacy of solar power, reducing precipitation, and unknown effects because they cannot be tested (a test of applying stratospheric sulfate aerosols is actually doing it in the actual environment) (see Robock 2009). The ecological crises often require us to act less (Epimethean inaction), not more (Promethean action). In the old saying often attributed to Gandhi or Mother Seton, "Live simply so that others may simply live."

Another Little Reflection

At once, three sparrows flee from the birdfeeder, detecting an unseen cue. Plop. And a second later, "jay, jay!" The immodest blue jay brooks no company while picking through her sunflower seeds.

Listening, I plan next year's garden. In my living room, I leaf through the Whole Seed Catalog. November's planning is better than October's uprooting. Small, capped-off shrub stumps infested the backyard garden area. Removing five, I reclaimed my newly rented yard. In springtime, I'll remove two in the front if I hope to plant a rose shrub or two—I imagine Mme. Hardy or Blanc de Vibert. But today, I dream of tomatoes and corn. There is "a time for every affair under the heavens" (Eccles. 3: 1 NABRE) and, in Pennsylvania, November is a time for planning.

Eating local eggs purchased at Nature's Pantry, with cheese from Berkey Creamery, and dried tomatoes from last month's North Atherton Farmers' Market, not much oil went into shipping today's breakfast. Birdsong streaming through the window has me connected in a way streaming music cannot. Sound and stomach interconnect in the experience of this soil. I slow down and feel today's soundscape, a present I did nothing to make. Today I live here, cultivating this place and myself in it.

No None

No none can place St. Francis's bell
  beneath the roar of engines bold
No none can hear the carolers sing
  in city streets where fears take hold
Strong winds defaced our home before
  and people's work rebuilt it
But people tear down mountains old
  that none, no none can ever replace
  no none can ever replace
Staying undone, a virtue becomes,
  when roots in soil take hold

Notes

1 From *Daybreak* (Baez 1968: 135)
2 This is the position of Catholic ecotheology, discussed in Chapter 5, "Spiritual Praxis." "The ideal ecological process, on the other hand, must be a complete process, one that includes the physical and the spiritual as well as the human dimensions of reality" (Berry 2009: 2).

References

Baez, Joan. 1968. *Daybreak*. New York: Dial Press.

Bates, Vincent C. 2013. "Drawing from Rural Ideals for Sustainable School Music." *Action, Criticism, and Theory for Music Education* 12, no. 1 (April): 24–46.

Berry, Thomas. 2009. "Spiritual Tradition and the Human Community." [Originally published in 1987.] In *The Christian Future and the Fate of Earth*, edited by Mary Evelyn Tucker and John Grim, 1–7. Maryknoll, NY: Orbis.

Bowers, C. A. 2007. *The False Promises of Constructivist Theories of Learning: A Global and Ecological Critique*. New York: Peter Lang.

Boykoff, Maxwell T. 2013. "Public Enemy No. 1? Understanding Media Representations of Outlier Views on Climate Change." *American Behavioral Scientist* 57, no. 6: 796–817. DOI: 10.1177/0002764213476846.

Brown, Steven E. F. 2014. "Suburban Pollution Cancels Savings from Dense Urban Cities." *San Francisco Business Times*, January 6. Online: http://www.bizjournals.com/sanfrancisco/news/2014/01/06/suburban-pollution-cancels-city-savings.html.

Fox, Matthew. 2012. *Hildegard of Bingen: A Saint for Our Times*. Vancouver: Namaste.

Gillis, Justin. 2014. "Group Earns Oil Income despite Pledge on Drilling." *New York Times*, Science, August 3. Online: https://www.nytimes.com/2014/08/04/science/group-earns-oil-income-despite-pledge-on-drilling.html.

Illich, Ivan. 1968. "To Hell with Good Intentions." Address to the Conference on InterAmerican Student Projects (CIASP), Cuernavaca, Mexico, April 20. Online: http://www.swaraj.org/illich_hell.htm.

——. 1990. "Declaration on Soil: A Joint Statement, Drafted in Hebenshausen, Germany, December 6, 1990, in Collaboration with Sigmar Groeneveld, Lee Hoinacki and Other Friends." Online: http://davidtinapple.com/illich/1990_declaraion_soil.PDF.

Koza, Julia Eklund. 2006. "'Save the Music'? Toward Culturally Relevant, Joyful, and Sustainable School Music." *Philosophy of Music Education Review* 14, no. 1 (Spring): 23–38.

McCall, Leslie. 2005. "The Complexity of Intersectionality." *Signs* 30, no. 3 (Spring): 1771–1800.

Morton, Charlene A. 2012. "Music Education for 'All My Relations.'" In *The Oxford Handbook of Philosophy in Music Education*, edited by Wayne D. Bowman and Ana Lucía Frega, 472–491. New York: Oxford University Press.

Palmer, Parker J. 2004. *A Hidden Wholeness: The Journal toward an Undivided Life*. San Francisco, CA: Jossey-Bass.

——. 2008. *The Promise of Paradox: A Celebration of Contradictions in the Christian Life*. San Francisco, CA: Jossey-Bass.

Prakash, Madhu Suri, and Gustavo Esteva. 2008. *Escaping Education: Living as Learning within Grassroots Cultures*, 2nd edition. New York: Peter Lang.

Robock, Alan. 2009. "The Many Problems with Geoengineering Using Stratospheric Aerosols." Paper presented at American Physical Society Meeting, Denver, CO, May. Online: https://www.aps.org/units/fps/meetings/april2009/upload/geoengineering-robock09.PDF.

Scruton, Roger. 2012. *How to Think Seriously about the Planet: The Case for an Environmental Conservatism*. New York: Oxford University Press.

Seeger, Anthony. 2016. "Natural Species, Sounds, and Humans in Lowland South America: The Kīsêdjê/Suyá, Their World, and the Nature of Their Musical Experience." In *Current Directions in Ecomusicology: Music, Culture, Nature*, edited by Aaron S. Allen and Kevin Dawe, 89–98. New York: Routledge.

Shevock, Daniel J. 2015. "The Possibility of Eco-Literate Music Pedagogy." *TOPICS for Music Education Praxis* 2015, no. 1: 1–23.

Index